AF564414

# Role of Agrometeorology Under Climate Change

NIPA® GENX ELECTRONIC RESOURCES & SOLUTIONS P. LTD.
New Delhi-110 034

## About the Author

**Dr. Sunil Kumar** is presently working as Assistant Professor-cum-Junior Scientist in the Department of Agronomy, BAU Sabour. He has completed his B.Sc. (Ag) from B.H.U., Varanasi and M.Sc. (Ag) in Agrometeorology from Govind Ballabh Pant University of Agriculture and Technology, Pantnagar. He has completed his Ph.D. in Agrometeorology and Environmental Sciences from SHUATS, Prayagraj, U.P. He has availed ICAR-JRF in Physical Sciences for pursuing his M.Sc. (Ag) and CSIR-JRF in Earth Atmosphere, Ocean and Planetary Sciences. He was given Young Scientist Award for outstanding contribution in 2017 by Society for Agriculture Innovation and Development (SAID), Ranchi (Jharkhand), India. He received Excellence in Teaching Award for outstanding contribution in Teaching in 2018. He has been actively involved in teaching, research and extension at BAU Sabour since 2012. He had published several research papers in national and international journals of repute. He had also a significant contribution in the field of Aerosol and its impact on Crop productivity, Crop Simulation, Agromet Advisory Services, Remote Sensing and GIS at Bihar Agricultural University, Sabour

# Role of Agrometeorology Under Climate Change

**Sunil Kumar**
Assistant Professor-cum-Junior Scientist
Department of Agronomy
Bihar Agricultural University
Sabour, Bihar

**NIPA® GENX ELECTRONIC RESOURCES & SOLUTIONS P. LTD.**
New Delhi-110 034

**NIPA® GENX ELECTRONIC RESOURCES & SOLUTIONS P. LTD.**

101,103, Vikas Surya Plaza, CU Block
L.S.C.Market, Pitam Pura, New Delhi-110 034
Ph : +91 11 27341616, 27341717, 27341718
E-mail:newindiapublishingagency@gmail.com
www: www.nipabooks.com

For customer assistance, please contact
Phone: + 91-11-27 34 17 17
Fax: + 91-11-27 34 16 16
E-Mail: feedbacks@nipabooks.com



ISBN: 978-81-19072-28-6



Composed and Designed by NIPA.

# Preface

Weather is the most important entity in agricultural production. Sustainable agricultural production is dependent to a large extent on precise knowledge of the weather resources. Precise measurements of weather parameters are required to understand the proper interaction in relation to crop growth and development.

This book offers a complete primer, covering the changes in water resources due to climate change and the end-to-end process of forecast production and bringing together a description of all the relevant aspects together in a single volume with plenty of explanation and examples of current field situation. The book covers the whole process of forecast production, from understanding the nature of the forecasting problem, gathering the observational data with which to initialise and verify forecasts, and then interpreting the output and putting it into a form which is relevant to farmers.

This is a rapidly developing field, with a lot of variations in practices between different forecasting centres. The author has tried to be as generic as possible when describing aspects of weather forecast procedure. The human forecaster still has a big part to play in producing weather forecasts and this is described along with the issue of forecast verification – how forecast centres measure their own performance and improve upon it.

This book provides description of weather forecasting process and uses a very practical approach, using real life case-studies to contextualize information. Discusses the latest advances in the area, including ensemble methods, monthly to seasonal range prediction and use of forecast in farmers' fields at village level. The book also describes how forecast on annual, seasonal and weekly weather data of a region is helpful to design different agricultural operations like field preparation, sowing, irrigation, fertilizer application as well as overall crop planning and livestock management.

Advanced undergraduates and postgraduate students will use this book to understand how the theory comes together in the day-to-day applications of weather forecast production. In addition, professional weather forecasting practitioners, professional users of weather forecasts and trainers will all find this book very helpful.

**Sunil Kumar**

# Contents

# List of Figures

# List of Figures

# List of Tables

# List of Tables

# 1

# Water Resources and Climate Change

## 1.1 Introduction

Climate change is evident from the observation of increase in global average air and ocean temperatures, precipitation and extreme rainfall, widespread melting of ice, storms / storm surges flooding and rising global mean sea level, as recorded in the Fourth Assessment Report of IPCC. Eleven of the last twelve years (1995-2006) rank among the twelve warmest years in the instrumental record of global surface temperature (since 1850). The 100-year linear trend (1906-2005) of 0.74 [0.56 to 0.92]°C is larger than the corresponding trend of 0.6 [0.4 to 0.8]°C (1901-2000) given in the Third Assessment Report (TAR). The linear warming trend over the 50 years 1956-2005 (0.13 [0.10 to 0.16]°C per decade) is nearly twice that for the 100 years 1906-2005. Global average sea level rose at an average rate of 1.8 [1.3 to 2.3] mm per year during 1961 to 2003 and at an average rate of about 3.1 [2.4 to 3.8] mm per year from 1993 to 2003. There is observational evidence of an increase in intense tropical cyclone activity in the North Atlantic since 1970, and suggestions of increased intense tropical cyclonic activity in some other regions where concerns over data quality are greater. Observed decreases in snow and ice extent are also consistent with warming. Globally, the area affected by drought has increased since 1970s (IPCC, 2008). In the 21$^{st}$ Century, according to Fourth Assessment Report of IPCC, during next two decades a warming of about 0.2°C per decade is projected for a range of SRES emission scenarios. Even if the concentrations of all GHGs and aerosols had been kept constant at year 2000 levels, a further warming of about 0.1°C per decade would be expected. Afterwards temperature projections increasingly depend on specific emission scenarios. In case of temperature rise, the best estimate for a low scenario is 1.8°C and the best estimate for high scenario is 4°C.

In India, almost 67% of the glaciers in the Himalayan mountain ranges have retreated in the past decades. Available records suggest that Gangotri glacier is retreating about 30 meters per year. A warming is likely to increase the melting far more rapidly than the accumulation. Past 200 years of instrumental observations indicate that the summer monsoon rainfall has undergone multi-decadal epochal variations in terms of the frequencies of droughts/floods (i.e., alternating 20-30 year periods of more and less frequent droughts), however, on a smaller space scale, there are areas showing both increasing (e.g. west coast) and decreasing (e.g. east and central India) long term trends in monsoon rainfall and sharp decrease in rainy days (Goswami *et al.*, 2006; Rajeevan *et al.*, 2006; Ramesh and Goswami, 2007). Figure 1.1 shows the changes in the frequency of extreme rainfall over central India. Climate change is a global problem and India will feel the heat due to its unique geophysical and hydro climatic conditions. Presently, about 68% area is liable to drought, 8% area is prone to cyclone and 40 million hectare area ($1/8^{th}$ of total area) is prone to flood. In the decade 1990-2000, an average of about 4344 people lost their lives and about 30 million people were affected by disasters every year. In 2006 at global level, the most significant disasters in terms of economic damage was the flood in India i.e. US$3.39 billion (0.29% of the previous year GDP) and around 40 million were victims. The reported natural disasters and number of people killed were 21 and 1521 respectively during 2006 in India (CRED, 2007). Nearly 700 million rural people in India directly depend on climate-sensitive sectors (agriculture, forests and fisheries) and natural resources (water, biodiversity, mangroves, coastal zones and grasslands) for their livelihood. Under changing climate, food security of the country might come under threat. In addition, the adaptive capacity of dry land farmers, forest and coastal community is low. Climate change is likely to impact all the natural ecosystems as well as health (e.g. malaria) and socio-economic systems. Presently, more than 45% of the average annual rainfall including snowfall in the country is going waste by natural runoff to sea. Artificial recharge/rain water harvesting scheme is now being implemented in the country to minimize this runoff loss based on present rainfall scenarios over the country. However, for the success of this scheme, we need to focus on how the possible climate change will affect the intensity, spatial and temporal variability of the rainfall, evaporation rates and temperature in different agro-climatic regions and river basins of India.

Ground water has been the mainstay for meeting the domestic water needs of more than 80% of rural and 50% of urban population, besides fulfilling the irrigation needs of around 50% of irrigated agriculture. The impact of rainfall variation on the region's ground water resources is not well understood, even though groundwater forms about half of the region's water supply. This is largely due to the complex interactions among land use, aquifer properties, antecedent water table levels and the actual timing and intensity of individual rainfall events. In cases where the aquifer systems are saturated, reductions in rainfall may not have an immediate effect on water tables, but a reduction in rainfall in other conditions below a critical level could eliminate all infiltration beyond the vegetation root zone. As a gross approximation, recharge to the water table aquifer (from which most of our ground water is derived) might be expected to respond in a similar way as runoff to decreasing rainfall, but this remains a largely unproven assertion. There have been very few researches on the potential effects of climate change. Eltahir and Yeh (1999) assessed the asymmetric response of aquifer water level to floods and droughts. They reported that the drought left a significantly more persistent signature in the aquifer water level than the corresponding signature of the flood. To examine the relative importance of climate on ground water level variation, Chen *et al.* (2004) used cross-correlation analysis between historical climate records and ground water levels. Their results showed that the annual precipitation explained the variations in ground water levels significantly. Van der Kamp and Maathuis (1991) investigated the annual fluctuation of ground water levels as a result of loading by surface moisture considering both the theoretical aspects of aquifer characteristics and empirical data. They observed that the relatively poor correlation between the climatic parameters and the ground water levels was due to the distance of the observation wells from the climate stations. To link climate variables with ground water levels, the weather station should exist in the re-charge zone of the observation well (Van der Kamp and Maathuis, 1991; Chen *et al.,* 2002). But, for a large-scale ground water monitoring network this may not be possible. However, the ground water level data itself provide a direct means of measuring the overall impacts of both natural and anthropogenic changes to ground water resources. Although the ground water monitoring networks have existed for several years, very little research has been carried out internationally to interpret the water table and quality trends. Broers and Grift (2004) studied the ground water quality trends due to anthropogenic-induced changes in agricultural practices. In future water resources will come under increasing pressure in Indian subcontinent due to the changing climate. The climate affects the demand for water as well as the supply and quality. Particularly, in arid and semi-arid regions of India any shortfall in water supply multiplied with climate change will enhance competition for water use for a wide range of economic, social and environmental applications. Assessing the potential socioeconomic impacts of climate change involves comparing two future scenarios, one with and one without climate change. Uncertainties involved in such an assessment include: (1) the timing, magnitude and nature of climate change; (2) the ability of ecosystems to adopt either naturally or through managed intervention to the change; (3) future increase in population and economic activities and the impacts on natural resources systems; and (4) how society adapts through the normal responses of individual and businessman and through policy changes that offer the opportunities and incentives to respond. The uncertainties, the long times involved and the potential for catastrophic and irreversible impacts on natural resources systems raise questions as how to evaluate climate impacts and investments and other policies that would affect or be affected by changes in the climate. In view of the above, an attempt has been made in this study to give a brief resume of possible impact of climate change on India's surface and ground water resources.

## 1.2 India's Rainfall, Population, Food and Fresh Water Needs

Long period average annual rainfall over the contiguous Indian area is about 117 cm; however, this rainfall is highly variable both in time and in space. Almost 75% (88 cm ± 10 SD) of the long average annual rainfall comes down in the four months of June to September (SW monsoon). The heaviest rains of the order of 200-400 cm or even more occur over north-east India and along the Western Ghats of the peninsular India (Fig. 1.1). Largely, the annual average rainfall over the northern Indo-Gangetic plains running parallel to the foot hills of the Himalayas varies from about 150 cm in the east to 50 cm in the west. Over the central parts of the India and northern half of the peninsular India, it varies from 150 cm in the eastern half to about 50 cm on the lee side of the Western Ghats. In the Southern half of the Indian peninsula, average annual rainfall varies from 100 cm to 75 cm as we go from east to west. On the other hand, some regions in the extreme western part of the country, such as western Rajasthan, receive average annual rainfall of about 15 cm or even less. There are considerable intra-seasonal and inter-seasonal variations as well.

The summer monsoon rainfall oscillates between active spells with good monsoon and weak spells or the breaks in the monsoon rains. The year-to-year variability in monsoon rainfall (Fig. 1.2) leads to extreme hydrological events (large scale drought and floods) resulting in serious reduction in agricultural output and affecting the vast population and the national economy. A normal monsoon with an evenly distributed rainfall throughout the country is a bonanza, while an extreme event of flood or drought over the entire country or a smaller region constitutes a natural hazard. Hence the variation in seasonal monsoon rainfall may be considered a measure to examine climate variability/ change over the Indian monsoon domain in the context of the global warming.

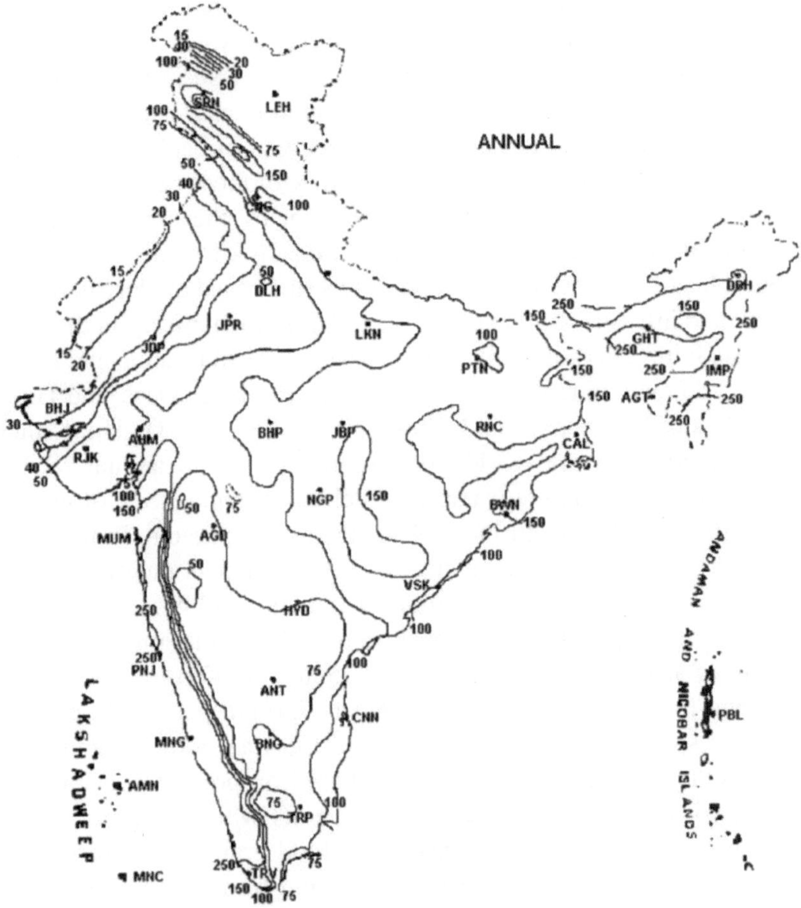

Fig. 1.1: Normal annual rainfall (cm). *Source*: IMD.

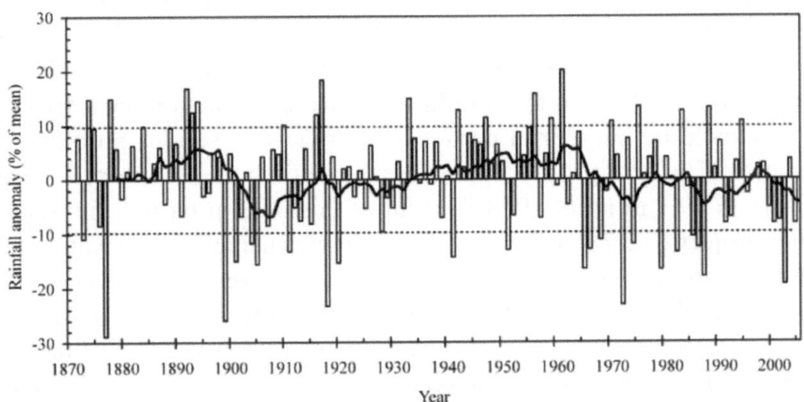


**Fig. 1.2:** All-India summer monsoon rainfall anomalies (1871–2004). Dark line shows 10-year moving average. (*Source*: Rajesh Kumar Mall, 2016)

Droughts, floods and desertification are directly connected with monsoon/ rainfall patterns, ocean circulation and soil moisture and water availability. As discussed above the problems of Indian rainfall are diverse in terms of both geographical distribution and seasonality, and spread over a period of years. There are large variations in the total rainfall received in each geographical division, causing both droughts and floods. The fury of these natural disasters has arguably been more intense and more frequent by the abuse of nature and degradation of environment. The adverse impacts of these two natural disasters cannot be assessed merely in economic terms based on destruction of crops, property and infrastructure because the toll of human misery in the form of death, disease, injury, loss of employment, psychological trauma, and above all the set-back to development are too difficult to evaluate (Dash and Hunt, 2007; Attwood, 2005; Prabhakar and Shaw, 2008; Revi, 2008). The United Nation has estimated that the world population grew at an annual rate of 1.4% during 1990-2000. China registered a much lower annual rate of growth (1.0%) along with USA (0.9%) during 1990-2000, as compared to India (1.9% during 1991-2001). It is investigated that if the National Population Policy (NPP) is fully implemented, the population of India should be 1,107 million by 2010 (Census of India, 2001). However, country's population is expected to reach a level of around 1,390 million by 2025 and 1,700 million by 2050. In India, average food consumption at present is 550 gm per capita per day whereas the corresponding figures in China and USA are 980 gm and 2850 gm, respectively. Present annual requirement based on present consumption level (550 gm) for the country is about 210 M tons which is almost equal to the current production. While the area under food grain, for instance, fell from 126.67 m ha to 123.06 m ha during the period from 1980-81 to 1999-2000, the production registered as increase from 129.59 M tons to 209 M tons during that period. The food grain production looked quite impressive in 1999-2000, which is more than 4 times the production of 50.82 M tons in 1950-51 (Directorate of Economics and Statistics, Ministry of Agriculture, New Delhi, India). It is feared that the fast increasing demand in next two or three decades could be quite grim particularly in view of serious problem of soil degradation. The total gross irrigated area has nearly trebled from 22.6 m ha in 1950-51 to 99.1 m ha in 1998-99. Out of this 34.3 m ha is from major and medium projects, 12.7 m ha from minor schemes using surface water and 52.2 m ha from groundwater. As against the national average of 38% of the total cropped area being irrigated, Punjab had the distinction of achieving highest-level irrigation (92%), followed by Haryana (79%) and Uttar Pradesh (66%) (CWC, 2002). The per capita average annual freshwater availability has reduced from 5177 cubic meters in 1951 to about 1820 cubic meters in 2001 and is estimated to further come down to 1341 cubic meters in 2025 and

1140 cubic meters (projected) in 2050. This clearly indicates the 'two sided' effect on water resources - the rise in population will increase the demand for water leading to faster withdrawal of water and this in turn would reduce the recharging time of the water tables. As a result, availability of water is bound to reach critical levels sooner or later.

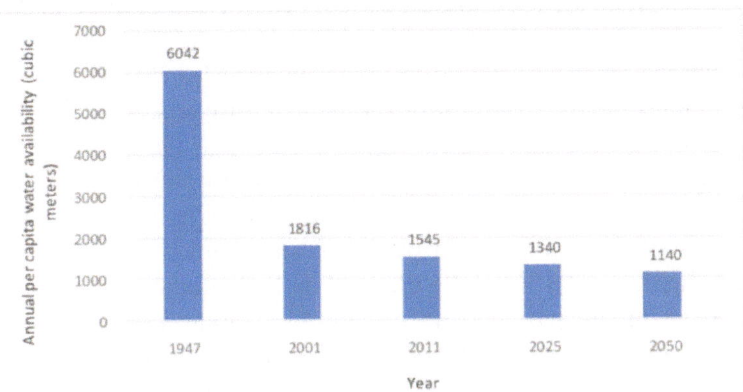


**Fig. 1.3:** Annual per capita water availability decrease pattern in India from 1947-2050. (*Source*: Vincent *et al.*, 2017)

## 1.3 Water Resources

### 1.3.1 Surface water resources

India has a large and intricate network of river systems of which the most prominent are the Himalayan rivers system draining the major plains of the country. Apart from this, numerous water bodies present in the subcontinent make it one of the wettest places in the world after South America. The annual precipitation including snowfall, which is the main source of the water in the country, is estimated to be of the order of 4000 billion cubic metres (Table 1.1). The Resource potential of the country, which occurs as natural runoff in the rivers is about 1869 cu. km. as per the basin wise latest estimates of Central Water Commission, considering both surface and ground water as one system. Ganga-Brahmaputra-Meghna system is the major contributor to total water resources potential of the country. Its share is about 60% in total water resources potential of the various rivers. Again about 40% of utilisable surface water resources are presently in Ganga-Brhamputra-Meghna system. In majority of river basins, present utilisation is significantly high and is in the range of 50 – 95% of utilisable surface resources. But in the rivers such as Narmada and Mahanadi the utilisation is quite low. The corresponding values for these basins are 23% and 34%, respectively. Some of the studies carried out in the Indian Himalayas clearly point out on an increase in glacial melt (Kumar *et al.*, 2007). For instance Baspa basin of Himachal Pradesh has shown an increase in the winter stream flow by 75% as compared to 1966. It is estimated that Himalayan Mountains cover surface area of permanent snow and ice in the region is about 97,020 km$^2$ with 12,930 km$^2$ volumes. In these mountains, it is estimated that 10 to 20% of the total surface area is covered by glaciers while an additional area ranging from 30 to 40% has seasonal snow cover (Upadhyay, 1995; Bahadur, 1999). These glaciers provide the snow and the glacial-melt waters keep our rivers perennial throughout the year. Bahadur (1999) reported that a very conservative estimate gives at least 500 km/year from snow and ice melt water contributions to Himalayan streams, while Afford (1992) reports about 515 km/year from the upper mountains. The most useful facet of glacial runoff is the fact that glaciers release more water in a drought year and less water in a flood year and thus ensuring water supply even during the lean years. The snow line and glacier boundaries are sensitive to changes in climatic conditions. Almost 67% of the glaciers in the Himalayan mountain ranges have retreated in the past decade (Ageta & Kadota, 1992; Yamada *et al.,* 1996). The mean equilibrium line altitude at which snow accumulation is equal to snow ablation for glacier is estimated to be about 50-80 meters higher relative to the altitude during the first half of the 19$^{th}$ Century (Pender, 1995). Available records suggest that Gangotri glacier is retreating about 28 meters per year. A warming is likely to increase the melting far more rapidly than the accumulation. Glacial melt is expected to increase under changed climate conditions, which would lead to increased summer flows in some river systems for few decades, followed by a reduction in flow as the glaciers disappear (IPCC, 1998). On the average, the area actually affected by floods every year in India is of the order of 10 million hectares of which about half is crop land. Rashtriya Barh Ayog (RBA) constituted by the Government of India in 1976 carried out an extensive analysis to estimate the flood-affected area in the country. RBA in its report has assessed the area liable to floods as 40 million hectares, which is nearly 1/8 of the country's area (Mall *et al.,* 2006).

**Table 1.1:** Water resources of India

| **Annual precipitation** | **4000 b.cu.m.** |
|---|---|
| Available water resources | 1869 |
| Utilizable | 1122 |
| Surface water (storage and diversion) | 690 |
| Groundwater (replenishable) | 432 |
| Present utilization | 605 |
| (Surface water 63%, groundwater 37%) | |
| Irrigation | 501 |
| Domestic | 30 |
| Industry, energy and other uses | 74 |

(*Source*: Hand book of Agriculture, ICAR)

According to CWC (2002), as many as 99 districts spread over 14 states were identified as drought prone districts in the country. Most of the drought prone areas are concentrated in the states of Rajasthan, Karnataka, Andhra Pradesh, Maharashtra and Gujarat. Human factors that influence drought include demand for water through population growth and agricultural practices, and modification of land use that directly influences the storage conditions and hydrological response of catchments and thus its vulnerability to drought. As pressures on water resources grow, vulnerability to meteorological drought has been increased (WMO, 2002). Sinha Ray and Shewale (2001) used rainfall data from 1875 to 1998 and give the percentage area of the country affected by moderate and severe drought. It may be noted that during the complete 124 years period there were three occasions i.e. 1877, 1899 and 1918 when percentage of the country affected by drought was more than 60%. It may be noted that during last few years there was no occasion when the percentage area of the country affected by drought was more than 50%. It also confirms the finding of Sen and Sinha Ray (1997), which showed a decreasing trend in the area affected by drought in the country. In 124 years, the probability of occurrence of drought was found maximum in Rajasthan (25%), Saurastra & Kutch (23%), followed by Jammu & Kashmir (21%) and Gujarat (21%) region. The drought of 1987 in various parts of the country was of "unprecedented intensity" resulting in serious crop damages and an alarming scarcity of drinking water. Only 12 of 35 meteorological sub-divisions in the country had received normal rains (Mall *et al.*, 2006, 2007). With the decreased rainfall contributing to drought, the water levels in major reservoirs in the country, meant for agricultural irrigation purposes and hydroelectric power, naturally declined.

### 1.3.2 Groundwater resources

Ground water plays a very important role in meeting the ever increasing demands of agriculture, industry and domestic sectors. India is a vast country having diversified geological, climatological and topographic set-up, giving rise to divergent ground water situations in different parts of the country. The prevalent rock formations, ranging in age from Achaean to recent, which control occurrence and movement of groundwater, are widely varied in composition and structure. Similarly, not too insignificant, are the variations of land forms, from the rugged mountainous terrains of the Himalayas, Eastern and Western Ghats to the flat alluvial plains of the river valleys and coastal tracts, and the Aeolian deserts of Rajasthan. The rainfall pattern also shows similar region wise variations. The topography and rainfall virtually control runoff and ground water recharge is controlled by these two in addition to composition and geometry of the aquifer.

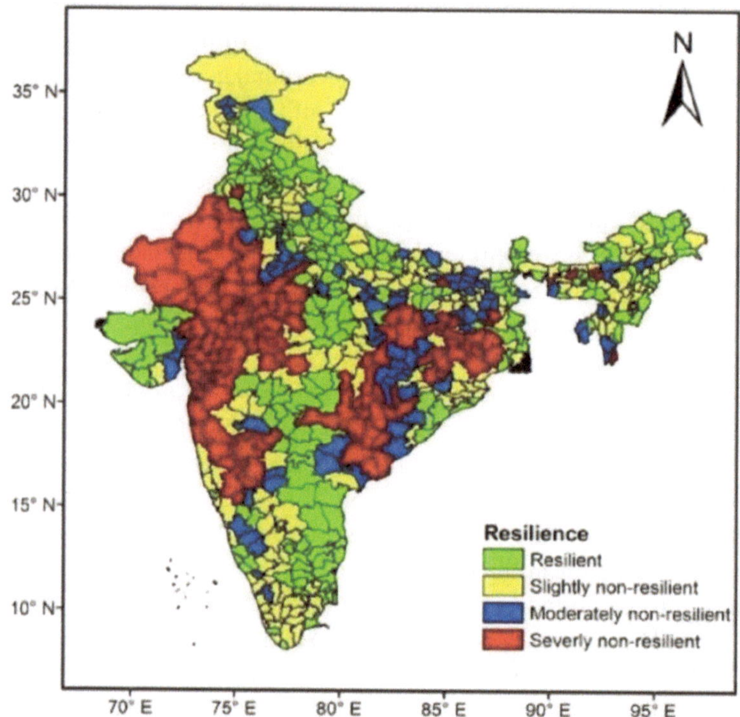


**Fig. 1.4:** Drought-Prone Districts of India. (*Source*: Indian Institute of Technology, Indore and Guwahati)

**Table 1.2:** Groundwater resource of India (b.cu.m/yr). (*Source*: CGWB)

| State/Union Territory | Total replenishable groundwater resource |
|---|---|
| Andhra Pradesh | 35.29 |
| Arunachal Pradesh | 1.44 |
| Assam | 24.72 |
| Bihar | 26.99 |
| Chhattisgarh | 16.07 |
| Delhi | 0.29 |
| Goa | 0.22 |
| Gujarat | 20.38 |
| Haryana | 8.53 |
| Himachal Pradesh | 0.37 |
| Jammu and Kashmir | 4.43 |
| Jharkhand | 6.53 |
| Karnataka | 16.19 |
| Kerala | 7.90 |

| | |
|---|---|
| Madhya Pradesh | 34.82 |
| Maharashtra | 37.87 |
| Manipur | 3.15 |
| Meghalaya | 0.54 |
| Mizoram | 1.40 |
| Nagaland | 0.72 |
| Orissa | 20.00 |
| Punjab | 18.66 |
| Rajasthan | 12.71 |
| Sikkim | 0.07 |
| Tamil Nadu | 26.39 |
| Tripura | 0.66 |
| Uttar Pradesh | 81.12 |
| Uttaranchal | 2.70 |
| West Bengal | 23.09 |
| Andaman and Nicobar Islands | 0.326 |
| Chandigarh | 0.030 |
| Dadar and Nagar Haveli | 0.042 |
| Daman and Diu | 0.013 |
| Lakshadweep | 0.002 |
| Pondicherry | 0.029 |

## 1.4 The Scarcity

The scarcity of water is understood in a very simple terminology of supply demand imbalances. Initially, the good olden days, generations have grown up with an idea of water as a free gift of nature, meaning thereby, it is available in abundance against the then demand situation. The challenge would be much more in coming years when the demand curve would shift upwards with relatively inelastic supply (demand growth rate would be higher than the supply growth rate) due to increased demand of agricultural, infrastructure development, uneconomic use, high population pressures etc. and here is the challenge of management of increased demand and low supply. Garg and Hassan (2007) pointed out that water scarcity is alarming and calls for urgent action before it becomes unmanageable. Das (2008) has suggested climatic changes to global warming will make water an increasingly scarce commodity in the coming years.

## 1.5 Scenario of Groundwater Resources

Groundwater is replenishable finite resource. Rainfall is the principle source of its recharge, in some areas though canal seepage and return flow from irrigation also contribute significantly to the groundwater recharge. Groundwater resources comprises of two parts namely dynamic, in the zone of water table fluctuation and static resource, blow this zone, which usually remains perennially saturated. As per the National Water Policy, 2002, the dynamic groundwater resource is essentially the exploitable quantity of groundwater, which is recharged annually, and is also termed as replenish able groundwater resource. The annual replenishable groundwater resource of the country is 433 billion cubic metres (bcm) and the net groundwater availability is 399 bcm after allocating 34 bcm for natural discharges during non-monsoon season. The annual groundwater draft for the year 2004 was 231 bcm, out of which 213 bcm is utilized for irrigation and 18 bcm is used for domestic and industrial purposes. Overall stage groundwater development is 58% (Chatterjee, 2009). However, 1615 assessment units fall under semi-critical (550), critical (226) and over Exploited (839) category, indicating that the groundwater resources in these areas is already being developed, more than that is being recharged. Further areas of the 30 assessment units are completely covered by saline groundwater. The rainfall contributes 67% of the country's annual replenishable groundwater resource (Chatterjee, 2009), indicating the dependence on rainfall for recharge of groundwater resources. The south west monsoon contributes 73% of the country's annual replenishable groundwater recharge, taking place during *kharif* period of cultivation. The stage of groundwater development is high in the states of Delhi, Haryana, Punjab and Rajasthan and Union Territories of Daman & Diu and Pondicherry, where the overall stage of groundwater development is more than 100%. Groundwater recharge is significantly high is the Indo-Gangetic-Brahmaputra alluvial belt, where the rainfall is plenty and thick piles of unconsolidated alluvial formations are conducive for recharge. Recharge per unit area (ha) in these regions varies from 0.28 to 1.35 m. The coastal alluvial belt also has relatively high recharge, in the range of 0.16 to 0.40 m. In the Western India, which have arid climate, the annual rechargeis only 0.10 m. Similar is the case with major part of the southern peninsular India (Chatterjee, 2009). Based on crop water requirement and availability of cultivable land, utilizable irrigation potential has been estimated as 64.05 million hectares (mha) excluding 6.4 mha kept as reserve for any eventuality. The irrigation potential created till March, 1997 is estimated as 45.73 mha (CGWB, 2002). During the past four decades, there has been a phenomenal increase in the growth of ground water abstraction structures due to implementation of technically viable schemes for development of the resource, backed by liberal funding from institutional finance agencies, improvement in availability of electric power and diesel, good quality seeds, fertilizers, government subsidies, etc. During the period 1951-97, the number of dug wells increased from 3.86 million to

10.50 million, shallow tube wells from 3000 to 6.74 million and public bore/ tube wells from negligible to 90,000. Electric pump sets have increased from negligible to 9.34 million and Diesel pumps from 66,000 to about 4.59 million (Chadha and Sharma, 2000). There has been steady increase in area irrigated by ground water from 6.5 M.ha in 1951 to 41.99 M.ha in 1997. The ultimate irrigation potential from groundwater is 64.05 million ha (mha) as per report on 3rd Census of Minor Irrigation Schemes (2005) as compared to 46 mha of land currently under groundwater irrigation, indicating further scope for developing groundwater in some area like eastern and north-eastern parts of the country (Planning Commission, 2007). During VIIIth Plan, 1.71 million dug wells, 1.67 million shallow tube wells and 114,000 deep tube wells are added (CGWB, 2002). Growing demands of water in agriculture, industrial and domestic sectors and groundwater development has brought problems of over-exploitation of the resource, continuously declining water levels, sea water ingress in coastal areas & ground water pollution in different parts of the country. The falling ground water levels in various parts of the country have threatened the sustainability of ground water resource, as water levels have gone deep beyond the economic lifts of pumping. The adverse effects of groundwater over development are continuous decline of groundwater level, drying of wells, fall in yield of wells, thus reduction in command, reversal of hydraulic gradient in the coastal and seawater intrusion, decline in base flow of streams, extra-cost to be incurred for deeper wells and deepening of existing wells due to decline in water level, increase in lifting cost of electricity and consumption of more electricity, decline in farm production and in term of income, pollution of groundwater and so many such related ecological issues (Das, 2008). Central Ground Water Board has established about 15000 network monitoring stations in the country to monitor the water level and its quality. The water level in the country in major part of the area is generally not showing any significant rise/ fall. However, significant decline in the levels of ground water have been observed in certain pockets of 289 districts in the States of Andhra Pradesh, Assam, Bihar, Chhattishgarh, NCT Delhi, Jharkhand, Gujarat, Haryana, Karnataka, Kerala, Madhya Pradesh, Maharashtra, Orissa, Punjab, Rajasthan, Tamil Nadu, Tripura, Uttar Pradesh and West Bengal. The groundwater in most of the areas in the country is fresh. Brackish ground water occurs in the arid zones of Rajasthan, close to coastal tracts in Saurashtra and Kutch, and in some zones in the east coast and certain parts of Punjab, Haryana, Western Uttar Pradesh etc., which are under extensive surface water irrigation. The fluoride levels in the ground water are considerably higher than the permissible limit in vast areas of Andhra Pradesh, Haryana and Rajasthan and in some parts of Punjab, Uttar Pradesh, M.P, Karnataka and Tamil Nadu. In the north-eastern regions, groundwater with iron content above the desirable limit occurs widely. Wide spread Arsenic contamination in West

Bengal Basin, covering eastern parts of West Bengal and in localized extent in Mid-Ganga basin is recorded (Saha, *et al.,* 2009). The Upper part of the new alluvial deposits underlying flood prone areas bordering Ganga River is affected by high incidence of arsenic contamination in groundwater. Pollution due to human and animal wastes and fertilizer application has resulted in high levels of nitrate and potassium in ground water in some parts of the country. Groundwater contamination in pockets of industrial zones is observed in localized areas. The over-exploitation of the coastal aquifers in the Saurashtra and Kutch regions of Gujarat has resulted in salinization of coastal aquifers. The excessive ground water withdrawal near the city of Chennai has led to seawater intrusion into coastal aquifers. The artificial recharge techniques can be utilised in improving the quality of ground water and to maintain the delicate fresh water-salt water interface (CGWB, 2002).

At present, available statistics on water demand shows that the agriculture sector is the largest consumer of water in India. About 83% of the available water is used for agriculture alone. The quantity of water required for agriculture has increased progressively through the years as more and more area was brought under irrigation. Since 1947, irrigated area in India has risen from 22.60 m ha to 80.76 m ha up to June 1997. The contribution of surface and groundwater resources for irrigation has played a significant role in India attaining self-sufficiency in food production during the past 3 decades and is likely to become more critical in future in the context of national food security. According to available estimates, the demand on water in this sector is projected to decrease to about 68% by the year 2050 though agriculture will remain the largest consumer. In order to meet this demand, augmentation of existing water resources by development of additional sources of water or conservation of the existing resources through impounding more water in the existing water bodies and its conjunctive use will be needed (Mall *et al.,* 2006). The depletion of groundwater and enhanced demand of water will have high impact on the sociocultural and economic fabric. The scarcity of water, might force people to make shift from agriculture to other alternative options. It would also force people to migrate to the cities (as already indicated in the studies conducted, that by 2050, 50% of the people will live in the urban areas) which would have other consequences. More pressure on urban infrastructure, cost of living, sustainable livelihood issues, housing, water sanitation, quality of life etc.

## 1.6 Observed Climate Change and its Impacts During the Past Century

Recently, Goswami *et al.* (2006) found that the frequency of occurrence as well as intensity of heavy and very-heavy rainfall events have highly significant increasing trends; low and moderate events have significant decreasing trend over Central India (Fig. 1.1). In India, several studies show that there is increasing trend in surface temperature i.e. 0.5 to 0.6°C during 1901-2005 and 0.05°C/decade year during the period 1901-2003, the recent period 1971-2003 has seen a relatively accelerated warming of 0.22°C/decade (Singhand Sontakke, 2002; Kothawale and Rupakumar, 2005; Mall *et al.,* 2006; Das & Hunt 2007), no significant trend in rainfall and/or decreasing/increasing trends in rainfall and sharp decrease in rainy days (Singh and Sontakke, 2002; Goswami *et al.,* 2006; Rajeevan *et al.,* 2006; Ramesh and Goswami, 2007). Singh and Sontakke (2002) found that the summer monsoon rainfall over western Indo-Gangetic Plain Region (IGPR) showed increasing trend (170 mm/100 years, significant at 1% level) from 1900, while over central IGPR it showed decreasing trend (5 mm/100 years, not significant) from 1939, and over eastern IGPR, decreasing trend (50 mm/100 years, not significant) during 1900-1984, and insignificant increasing trend (480 mm/100 years, not significant) was observed during 1984-1999. Broadly, it is inferred that there has been a west ward shift in rainfall activities over the IGPR. The year-to-year variability in monsoon rainfall leads to extreme hydrological events (large scale drought and floods) resulting in serious reduction in agricultural output and affecting the vast population and the national economy. Droughts, floods and desertification are directly connected with monsoon/rainfall patterns, ocean circulation and soil moisture and water availability. As discussed above the problems of Indian rainfall are diverse, in terms of both geographical distribution and seasonality, and spread over a period of years. There are large variations in the total rainfall received in each geographical division, causing both droughts and floods. The fury of these natural disasters has arguably been more intense and more frequent by the abuse of nature and degradation of environment. The adverse impacts of these two natural disasters cannot be assessed merely in economic terms based on destruction of crops, property and infrastructure because the toll of human misery in the form of death, disease, injury, loss of employment, psychological trauma, and above all the set-back to development are too difficult to evaluate (Dash and Hunt, 2007; Prabhakar and Shaw, 2008; Revi, 2008). Reddy *et al.* (2008) reported that during 1990 to 2004 Kosi River showed a significant shift of 3.5 km in north-western part, followed by central and north eastern parts of river with 2.5 km shift. Course change by the rivers is an environmental problem of serious concern in the Indo-Gangetic Plain Region (IGPR). At different times in the past different rivers changed their course a number of times. During the period 1731-1963, the course of the Kosi river (the sorrow of Bihar) has shifted westward by about 125 km, the courses of Ganga, Ghaghara and son at their confluence have shifted by 35 to 50 km since epic period ~ 1000 BC (Singh, 1971) and that of Indus and its tributaries by 10-30 km in the 1200 years in the same (Wilhelmy, 1967). Between 2500 BC and 500 BC the course of the Yamuna river shifted westward to join Indus and then east to join Ganga thrice (Raikes, 1968). Das and Radhakrishnan (1991) reported a rising trend in the sea level at Mumbai (Bombay) during 1940-86 and Chennai (madras) during 1910-1933, based on the annual means of tide gauge observations. Srivastava and Balakrishnan (1993) confirmed a rise of sea level by 8 cm with a corresponding fall in the pressure during 1901-1940. Unnikrishnan *et al.* (2006) estimated sea level rise at selected stations to be around 10cm per century. Such prediction has been made by analysing past tide gauge data. The estimation for sea level rise for four major cities like Mumbai, Kochi, Vishakhapatnam and Chennai were 0.78, 1.14, 0.75 and -0.65 mm/year. By above studies it is clear that the global warming threat is real and the consequences of the climate change phenomena are many, and alarming. The impact of future climatic change may be felt more severely in developing countries such as India whose economy is largely dependent on agriculture and is already under stress due to current population increase and associated demands for energy, fresh water and food. In spite of the uncertainties about the precise magnitude of climate change and its possible impacts particularly on regional scales, measures must be taken to prevent or minimize the causes of climate change and mitigate its adverse effects.

## 1.7. Projected Climatic Trend

Rupa Kumar *et al.* (2006) projected that warming is monotonously wide spread over the country, but there are substantial spatial differences in the projected rainfall changes. West central India shows maximum expected increase in rainfall. Extremes in maximum and minimum temperatures are also expected to increase in future, but the night temperatures are increasing faster than the day temperatures. Extreme precipitation shows substantial increases over a large area, particularly over the west coast of India and west central India. However, projection over north eastern region shows not satisfactory result and more in depth study is going on. Lal *et al.* (2001) estimated that $CO_2$ level will increase to 397-416 ppm by 2010s from the present $CO_2$ level of 371 ppm and this would further increase by 605-755 by 2070. They projected between 1 to 1.4°C and 2.23 to 2.87°C area-averaged annual mean warming by 2020 & 2050, respectively. Comparatively, increase in temperature is projected to be more in winter season than in summer. A large uncertainty is associated with projected winter rainfall than monsoon rainfall in 2050s. Moreover, the standard deviation of future projections of area averaged monsoon rainfall centred in 2050s is not significantly different relative to the present-day atmosphere implying thereby that the year-to-year variability in mean rainfall during the monsoon season may not significantly change in the future. More intense rainfall spells are, however, projected over the land regions of the Indian subcontinent in the future thus increasing the probability of extreme rainfall events in a warmer atmosphere. Rupa Kumar and Ashrit (2001) have projected 13% increase in monsoon or *kharif* season rainfall in India using ECHAM4 model, while HadmodCM2 suggests reduction in *kharif* rainfall by 6% in the greenhouse gas simulation. Both GCMs suggest an increase in annual mean temperature by more than 1°C (1.3°C in ECHAM4 and 1.7°C HadCM2). Rupa Kumar *et al.* (2003) concluded that future scenarios of increased greenhouse gas concentrations (GHG) indicate marked increase in both rainfall and temperature into the 21st century, particularly becoming conspicuous after the 2040s in India. Over the region south of 25°N (south of cities such as Udaipur, Khajuraho and Varanasi), the maximum temperature will increase by 2-4°C during 2050s. In the northern region, the increase in maximum temperature may exceed 4°C. This study also indicates a general increase in minimum temperature up to 4°C all over the country, which may however exceed over the southern peninsula, northeast India and some parts of Punjab, Haryana and Bihar. There is an overall decrease in number of rainy days over a major part of the country. This decrease is more in western and central part (by more than 15 days) while near the foothills of Himalayas (Uttarakhand) and in northeast India the number of rainy days may increase by 5-10 days. However, increase in GHG may lead to overall increase in the rainy days intensity by 1-4 mm/day except for small areas in the north-west India where the rainfall intensities decrease by 1 mm/day.

## 1.8 Impacts of Projected Climate Change on Water Resources

Table 1.4 shows the selective reports on impact of climate change on water resources during next century over India. The enhanced surface warming over the Indian subcontinent by the end of the next century would result in an increase in pre-monsoonal and monsoonal rainfall and no substantial change in winter rainfall over the central plains. This would result in an increase in the monsoonal and annual runoff in the central plains with no substantial change in winter runoff. They also indicated an increase in evaporation and soil wetness during the monsoon and on an annual basis (Lal and Chander, 1993). The regional effects of climate change on various components of the hydrological cycle, namely surface run-off, soil moisture, and evapotranspiration (ET) for three-drainage basins of central India is analysed and results indicated that the basin located in a comparatively drier region is more sensitive to climatic changes. The high probability of a significant effect of climate change on reservoir storage, especially for drier scenarios, necessitates the need of a further, more critical analysis of these effects. Chattopadhyay and Hulme (1997) calculated increases in potential evaporation across India from

GCM simulations of climate; they found that projected increases in potential evaporation were related largely to increases in the vapour pressure deficit resulting from higher temperature. The hydrologic sensitivity of the Kosi Basin to projected land-use and potential climate change scenarios has been analysed. It was found that runoff increase was higher than precipitation increase in all the potential climate change scenarios applying cotemporary temperature. The scenario of contemporary precipitation and a rise in temperature of 4°C caused a decrease in runoff by 2-8% depending upon the areas considered and model used (Sharma *et al.,* 2000a,b). It is also projected that soil moisture increase marginally by 15-20% over parts of Southern and Central India. This increase is confined to the monsoon months of June through September. During the rest of the year, there is either no change in soil moisture or a marginal decline possibly due to the increase in temperature leading to enhanced ET (Lal and Singh, 2001). Gosain *et al.* (2003, 2006) projected that the quantity of surface runoff due to climate change would vary across the river basins as well as sub basins in India. However, there is general reduction in the quantity of the available runoff. An increase in precipitation in the Mahanadi, Brahimini, Ganga, Godavari and Cauvery is projected under climate change scenario; however, the corresponding total runoff for all the basins does not increase. This may be due to increase in ET because of increased temperature or variation in the distribution of the rainfall. In the remaining basin, a decrease in precipitation was noticed. Sabarmati and Luni basin showed drastic decrease in precipitation and consequent decrease of total runoff to the tune of 2/3rd of the prevailing runoff. This may lead to severe drought conditions in future.

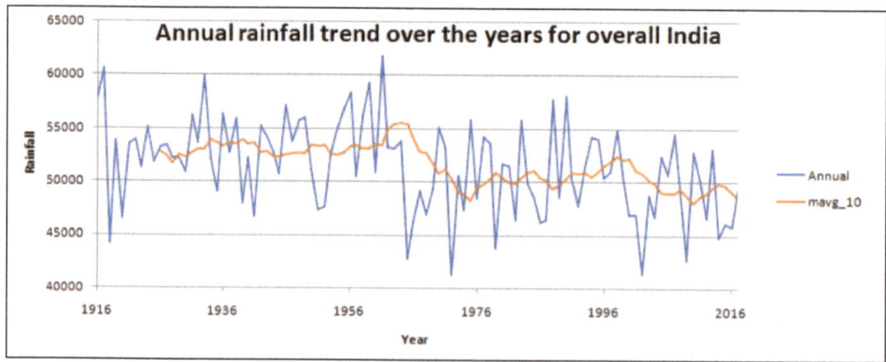


**Fig. 1.5:** Annual rainfall trend over India. (*Source*: Vighnesh Uday Tamse, 2019)

The analysis has revealed that climate change scenario may deteriorate the condition in terms of severity of droughts and intensity of floods in various parts of the country. There have been few more studies on climate change impacts on Indian water resources (Roy *et al.,* 2003; Chadha, 2003; Tangri,

2003). Singh *et al.* (2009) highlighted the assessment of the water resources in changing climate for relevant national and regional long-term development strategies. Goyal (2004) studied the sensitivity of ET to global warming for arid regions of Rajasthan and projected an increase of 14.8% of total ET demand with increase in temperature, however ET is less sensitive to increase in solar radiation, followed by wind speed in comparison to temperature. Increase in water vapour has a negative impact on ET (-4.3%). He concluded that a marginal increase in ET demand due to global warming would have a larger impact on resource poor, fragile arid zone ecosystem of Rajasthan.

**Table 1.3:** Impact of climate change on water resources.

| **Location** | **Impact** | **References** |
|---|---|---|
| Indian subcontinent | Increase in monsoonal and annual runoff in the central plains<br>• No substantial change in winter runoff.<br>• Increase in evaporation and soil wetness during the monsoon and on an annual basis. | Lal and Chander, 1993 |
| Orissa and West Bengal | One-meter sea levels rise would inundate 1700 km2 of prime agricultural land | IPCC, 1992 |
| Indian coastline | One-meter sea level rise on the Indian coastline is likely to affect a total area of 5763 km2, and put 7.1 million people at risk | JNU, 1993 |
| All India | Increases in potential evaporation across India | Chattopadhyay and Hulme, 1997 |
| Central India | Basin located in a comparatively drier region is moresensitive to climatic changes | Mehrotra, 1999 |
| Kosi Basin | Decrease in runoff by 2-8% | Sharma *et al.*, 2000, a,b |
| Southern and Central India | Soil moisture increase marginally by 15-20% in monsoon months | Lal and Singh, 2001 |
| Damodar basin | Decreased river flow | Roy *et al.*, 2003 |
| Rajasthan | An increase in ET | Goyal, 2004 |
| River basins of India | General reduction in the quantity of the available runoff, increase in Mahanadi and Brahmini basin | Gossai and Rao, 2006 |
| River basins in northwest & central India | Increase in heaviest rainfall and reduction in number of rainy days | Singh *et el.*, 2008 |

## 1.9 Ground Water and Climate Change

Problems in ground water management in India have potentially huge implications for global warming. The most optimistic assumption suggests that an average drop in ground water level by one meter would increase India's total carbon emissions by over 1%. More realistic assumption reflecting the area projected to be irrigated by groundwater in 2003, suggests that the increase in Carbon emission could be 4.8% for each meter drop in groundwater levels. Chadha (2003) recommended studying the aquifer geometry and establishing the saline fresh inter faces within 20 Km of the coastal area, the effect of glaciers melting on the recharge potential of the aquifer in the Ganga basin together with its effect on the trans-boundary aquifer system particularly of the arid and semi-arid regions. Panda *et al.* (2007) studied the influence of repeated droughts and increased anthropogenic pressure on the groundwater levels of Orissa during the period 1994–2003. Preliminary study showed that the groundwater levels of the network observation wells are very sensitive to the monsoon rainfall, and any irregularity in rainfall directly influences the groundwater levels. Due to drought in 2002, the groundwater level dropped significantly in the consolidated formation that covers 80% of the geographical area of Orissa. The fitted curves of both the annual and monsoon rainfall indicated a downward trend although four wet years were experienced during the study period. The effect of droughts and high temperature on groundwater levels should be counter balanced by the effect of flood, and over years it should remain stable.

## 1.10 Climate Change Policy

The Government of India is actively involved with climate change activities since long. India is a Part to the United Nations Framework Convention on Climate Change (UNFCC). The Eighth session of the Conference of Parties (COP-8) to the UN convention on Climate Change in 2002, New Delhi ended with a Delhi Declaration which successfully resolved the technical parameters necessary for the implementation of the Kyoto Protocol (1997). The Delhi declaration gave primacy for the implementation of the Clean Development Mechanism (CDM) in the climate change process. The National Clean Development Mechanism Authority is operational since December 2003 to support implementation of CDM projects. The Bali conference on climate change (December 2009) showed all the countries the way forward to the next phase of the campaign to control the planet's changing climate, the specific objective being to put a multilateral arrangement in place that will succeed the 1997 Kyoto Protocol of the UN convention on Climate Change, which will terminate in 2012. The Bangkok meeting (March 2008) was the beginning of the new process, which was continued in December 2009 at Copenhagen. To address the future challenges, in June 2007, the Government announced the constitution of a high level advisory group on climate change and prepared a 'National Action Plan on Climate Change' and that released by the Hon'ble Prime minister of India on June 30, 2008 (http:/pmindia.nic.in/Climate%20

Change_16.03.09.pdf); which is in line with the international commitments and contains eight missions on climate mitigation and adaptation (NAPCC, 2008). Now relevant ministries are preparing and submitting their respective plans to the Prime Minister's Climate Change Council. One of the missions "National Water Mission" will be mounted to ensure integrated water resources management helping to conserve water, minimize wastage and ensure more equitable distribution both across and within states. The mission will take into account the provisions of National Water Policy and develop a framework to optimize water uses and by increasing water use efficiency by 20% through regulatory mechanism and differential entitlements and pricing. It will seek to ensure that considerable share of water needs of urban areas are met through recycling of waste water, and ensuring that the water requirements of coastal cities with inadequate alternative sources of water are met through adoption of new and appropriate technologies such as low temperature desalination technologies that allow for the use of ocean water (Singh *et al.,* 2009).

Understanding about Climate change adaptation is still growing. In India, many regions and sectors, climate change impact assessment is yet to be completed. The projection of rainfall scenario over north-eastern region (Assam & Bihar) of the country is not satisfactory. Given that climate change assessment is still going on, there is need to develop and strengthen the climate change adaptation process which depend on more analysis of the resilience livelihood and the policies that impacts on livelihood and people capacities. Society already encounters large costs in adapting to climate extremes; climate change will only increase these costs. Several policies already exist to address adaptation to climate variability: however there is need to study these policies and their impact on communities' capacities to adapt to climate change in Assam and Bihar. Given our limited understanding of climate change, extending the range of adaptation strategies seems worthwhile. In this respect, special attention should be given to the following:

1. Alternative livelihood.
2. Capacity Building: to educate people about the effects of their activities on carbon –trapping and about possible responses to the effects of natural climatic variability and potential climate changes in the future.
3. Changes in land-use allocation, including the development of new plant species. Since most of the world's plant food comes from just 20 species, the potential of the majority of species is still to develop.
4. Food security policies and reduction of post-harvest losses. Post-harvest losses, due to deficient systems of storage and transport, amount to 50% or more in many states, which means that there is ample room for improvement.

5. Conversion to "controlled-environment agriculture" may invest billions of rupees in an annum.
6. Water resource management: Areas where large changes in rainfall regimes (e.g. increased frequency and severity of floods or drought), an improved and more environmentally sound infrastructure will be necessary and policies encouraging water conservation (e.g. pricing mechanisms) will have to be introduced.

Based on the this current study it may be mentioned that for long term adaptation from Climate Variability and Climate Change, several policy instruments are available at different levels and these instruments are functioning in real sense even at the remote village levels but need improved governance, productivity and accountability of the government machinery.

## 1.11 Conclusion

These studies are still in infancy and a lot more data in terms of field information is to be generated. This will also facilitate the appropriate validation of the simulation for the present scenarios. However, based on above studies it is clear that the global warming threat is real and the consequences of the climate change phenomena are many and alarming. The impact of future climatic change may be felt more severely in developing countries such as India whose economy is largely dependent on agriculture and is already under stress due to current population increase and associated demands for energy, fresh water and food. In spite of the uncertainties about the precise magnitude of climate change and its possible impacts particularly on regional scales, measures must be taken to anticipate, prevent or minimize the causes of climate change and mitigate its adverse effects. In addition, the uncertainty involved in predicting extreme flood and drought events by the models are large. Also there is no clear role of global warming in the variability of monsoon rainfall over India. Therefore, it is difficult, at this juncture, to convince the water planning and development agencies to incorporate the impact of climate change into their projects and water resources systems. However, given the potential adverse impacts on water resources that could bring about by climate change, it is worthwhile for the authority to conduct more in-depth studies and analyses to gauge the extent of problems that the country may face. Agricultural water demand, particularly for irrigation water, is considered more sensitive to climate change. A change in field-level climate may alter the need and for timing of irrigation: Increased dryness may lead to increase in demands, but demand could be reduced if soil moistures content rises at critical times of the year (IPCC, 2001). Doll and Sibert (2001) concluded that global net irrigation requirements would increase relative to the situation without climate change by 3.5 to 5% by 2025, and 6-8% by 2075. In a recent study by Saeed *et al.* (2009) signified the role of irrigation in effecting the local temperature, which in turn effects large-scale circulations and precipitation. From the above it can be concluded that Indian region is highly sensitive to climate change and demand for water from groundwater may increase if precipitation decreases and surface water inflows decrease, this leads to a decrease in discharge elsewhere. The elements / sectors currently at risk are likely to be highly vulnerable to climate change and variability and here existed uncertainties in dealing with vulnerabilities associated to climate change and variability. It is urgently required to intensify in-depth research work with following objectives:

- Strengthening observational data and data access network.
- Assess the recent experience in climate variability and extreme events, impacts of projected climate change and variability and associated hydrological events in India at river basin / aquifer level.
- How climate change might affect groundwater aquifers, including quality, recharge rates, and flow dynamics.
- In-depth study on groundwater recharge, which is so dependent on individual sustained rainfall events as well as on the changes in land use pattern. This study is long due.
- How sea level rise might affect the coastal groundwater regime.
- Determine vulnerability of regional water resources to climate change and identifying key risks and prioritizing adaptation responses.
- Community based water management, to conserve and augment recharge.
- Alternative cropping pattern; possible increase in irrigation demands, and its mismatch with water availability.

As discussed, climate change may have both direct and indirect effects on both recharge and discharge to an aquifer. Increased temperature may lead to higher potential evapotranspiration and increased water use demand. Therefore, an effective management of ground water resources requires an integrated approach in both planning and implementation of schemes. Different agencies related to water resources, climate, agriculture and other sectors should coordinate and bring out policies on scientific considerations for effective management of ground water resources in changing climate. Uncertainty in predictions from global climate models (GCMs) is currently limiting the ability of groundwater models to predict the impact of climate change on ground water. Greater consistency between GCMs and a finer resolution will allow better prediction of groundwater systems response to climate change that are subject to less uncertainty. The coupling of GCMs with hydrologic models is necessary due to advances in super-computer technology. A future increase in demand water is likely to have a much greater impact on groundwater than reduced recharge due to climate change. With increased scarcity of groundwater, the time has come when government and community should work together for an integrated management targeted towards providing water to all on a sustainable basis.

## References

Attwood, D.W., 2005. Big is Ugly? How Large-scale Institutions Prevent Famines in Western India, World Development, 33(12), 2067–2083.

Broers, H.P. and Grift, B., 2004. Regional monitoring of temporal changes in groundwater quality. J. Hydrol. 296, 192–220.

Central Ground Water Board, 2006. Dynamic ground water resources of India (as on March, 2004) New Delhi.

Chadha, D.K. 2003. Climate change and ground water resources of Deccan basalt and Ganga basin. In the Proceedingsof the NATCOM- V&A Workshop on Water Resources, Coastal Zones and Human Health, IIT Delhi, New Delhi, 27-28 June, 2003.

Chatterjee, R. and R.R. Purohit, 2009. Estimation of replenishable groundwater resources of India and theirstatus ofutilization, Current Science, 96(12), 1581-1591.

Chattopadhyary, N. and M. Hulme, 1997. Evaporation and potential evapo-transpiration in India under conditions of recent and future climate change. Agricultural and Forest Meteorology, 87, 55–73.

Chen, Z., Grasby, S. and Osadetz, K.G., 2004. Relation between climate variability and groundwater levels in the up NationalSymposium on Climate Change and Rainfed Agriculture, February, 18-20, 2010, CRIDA, Hyderabad, India 79 per carbonate aquifer, south Manitoba, Canada. J. Hydrol. 290, 43–62.

CRED, 2007. Annual disaster statistical review: Number and Trends 2006, Centre for Research on The Epidemiology of Disasters (CRED), School of Public Health, Catholic University of Louvain, Brussels, Belgium, p55.

CWC (Central Water Commission), 2002. Water and related statistics, New Delhi, 479 pp.

Das, S. 2008. Drinking Water and Food Security in Hard Rock Areas of India, Golden Jubilee Volume, Geological Society of India.

Das, S. and Anandhakumar, K.J., 2005. Management of Ground water in Coastal Orissa-International Journal of Ecology & Environmental Science, Vol. 31, No. 3, Special Issue (P285-297).

Das, S., 2008. Five decades of Ground water Development in India, Chanfging Geohydrlogical Scenario, Hard rock Terrain of Peninsular India, Geological Society of India, Golden Jubilee Volume (xiii–xxix).

Garg, N.K. and Hassan, Q. 2007. Alarming scarcity of water in India. Current science, 93(7), 932-941.

Gosain, A.K., and Rao, S. 2003. Impacts of climate on water sector. In: Shukla, P.R., Sharma, S.K., Ravindranath, N.H., Garg, A., and Bhattacharya, S. (eds.), Climate Change and India Vulnerability Assessment and Adaptation.Universities Press (India) Pvt. Ltd., Hyderabad, 462 pp

Gosain, A.K., Rao, Sandhya and Basuray, Debajit: 2006, Climate change impact assessment on hydrology of Indianriver basins, Current Science, 90(3), 346-353.

Goswami, B.N., V. Venugopal, D. Sengupta, M.S. Madhusoodanan, and P.K. Xavier, 2006. Increasing trend of extreme rain events over India in a warming environment, Science. 314, 1442–1445, doi:10.1126/science.1132027.

IPCC, 2007. "IPCC Fourth Assessment Report, Synthesis Report: Full Text." In: Encyclopedia of Earth, Stephen C.

Kothawale, D.R. and K. Rupa Kumar, 2005. On the recent changes in surface temperature trends over India. Geoph. Res. Letters, Vol. 32, L18714.

Kripalani R.H., Kulkarni, A., Sabade, S.S. & Khandekar, M.L. 2003. Indian monsoon variability in a global warmingscenario, Natural Hazards, 29(2), 189-206.

Kumar, R., Hasnain, S.I., Wagnon, P., Arnaud, Y, and Sharma P., 2007. Climate change signals detected through massbalance measurements on benchmark glacier, Himachal Pradesh, India. P 65-74. Climatic & anthropogenic impactson the variability of water resources. Technical Document in Hydrology, 80, UNESCO, Paris/UMR 5569, Hydro Science Montpellier, 2007.

Lal, M., Nozawa, T., Emori, S., Harasawa, H., Takahashi, K., Kimoto, M., Abe-Ouchi, A., Nakajima, T., Takemura,T. and Numaguti, A., 2001. Future climate change: Implications for Indian summer monsoon and its variability.Current Science, 81(9), 1196-1207.

Loaiciga, H.A., Maidment, D.R., Valdes, J.B., 2000. Climate change impacts in a regional karst aquifer, TX, USA. J. Hydrol. 227, 173–194.

Mall R.K., R. Bhatla and S.N. Pandey, 2007. "Water resources in India and impact of climate change", Jalvigyan Sameeksha (Min of Water Resources), 22, 157-176.

Mall, R.K., R. Singh, A. Gupta, Singh R.S., and L.S. Rathore, 2006. 'Water Resources and Climate Change: An Indianperspective, Current Science, 90(12), 1610-1626.

MoWR (Ministry of Water Resources), 2003. Vision for Integrated Water Resources Development and Management. Government of India, New Delhi, India, 20pp.

Panda, D.K, Mishra A., Jena S.K., James B.K and Kumar A., 2007. The influence of drought and anthropogenic effectson groundwater levels in Orissa, India Journal of Hydrology, 343, 140-153.

Planning Commission, 2007. Report of the Expert Group on Ground Water Management and Ownership.

Prabhakar, SVRK and Shaw, R, 2008. Climate Change adaptation implications for drought risk mitigation: a perspectivefor India, Climatic Change, 88, 113-130.

Ramesh, K.V. and Goswami, P., 2007. Reduction in temporal and spatial extent of the Indian summer monsoon.Geophy. Res. Letter, Vol. 34, L23704, doi:10.1029/2007 GL031613.

Reddy S.C., Rangaswamy M. and Jha C.S. 2008. Monitoring of spatio-Temporal Changes in part of Kosi River basin, Bihar, India using remote sensing and GIS, Research Journal of Environmental Sciences, 2(1), 58-62.

Revi, A., 2008. Climate change risk: an adaptation and mitigation agenda for Indian cities, Environment and Urbanization, 20(1), 207-229.

Richard A. Kerr, 2009. Northern India's Groundwater is Going, Going, Going, Science V 325(5942) p 798.

Rupakumar, K., Sahai, A.K., Kumar, Krishna, K., Patwardhan, S.K., Mishra P.K., Revadkar, J.V., Kamala, K., and Pant, G.B. 2006. High-resolution climate change scenarios for India for the 21st century, Current Science, 90(3), 334-345.

Saeed, F., S. Hagemann and D. Jacob, 2009. Impact of irrigation on the south Asian summer monsoon, Geophysical Research Letters, 36 ( L 20711).

Saha D., Sreehari Sarangam, Shailendra, N., and Kuldeep G. Bhartariya, 2009. Evaluation of hydrogeochemical processesin arsenic-contaminated alluvial aquifers in parts of Mid-Ganga Basin, Bihar, Eastern India, Environ Earth Sci., 0(0), 1-13.

Singh P., V. Kumar, Thomas V., and M. Arora, 2008. Changes in rainfall and relative humidity in river basins in North West and central India, Hydrological Process, 22, 2982-2992.

Singh R.D., M. Arora and Rakesh Kumar, 2009. Climate change impact on Water Resources in India, Mausam, (Diamond Jubilee volume, 2009), 91-100.

Singh, N. and Sontakke, N.A. 2002. On climatic fluctuations and environmental changes of the Indo-Gangetic plains,India. Climatic Change, 52, 287-313.

Sinha Ray, K.C. and She wale, M.P. 2001. Probability of occurrence of drought in various sub-divisions of India.Mausam, 52(3), 541-546.

Tangri, C.D., Impact of climate change on Himalayan glaciers. In the Proceedings of the NATCOM- V&A.Workshopon Water Resources, Coastal Zones and Human Health, IIT Delhi, New Delhi, 27-28 June, 2003.

WMO (World Meteorological Organization), 2002: Reducing vulnerability to weather and climate extreme, WMO–No. 936, Geneva, Switzerland, 36pp.

# 2

# Role of Weather Forecast in Indian Agriculture

## 2.1 Introduction

Agro-meteorological advisories relate the past, current and predicted future weather to agricultural activities. Behaviour of weather both in space and time, considerably affect the agricultural production. Year to year variations in crop production are generally depended to the abnormalities in the behaviours of weather. The mission of the Agromet Advisory Services (AAS) of NCMRWF is to maximise farmers' profits by decreasing losses due to weather calamities and also by increasing the timeliness of operations in the crop field. An ideal AAS bulletin helps to reduce environmental pollution through the optimum use of chemicals in the crop filed. An ideal AAS bulletin should contain farm management information to weather based agricultural operations and available to the farmers in real time.

Agricultural operations can be suitably planned and appropriate measures can be taken to offset adverse influences of weather of necessary forecasts be available. Thus, forecast of weather parameters plays a vital role in agricultural production. In most of the area, when weather occurrence is not sudden, availability and judicious use of Agro-meteorological information is advantageous for farm management. It helps in enhancing economic gains by suitable adjustment of the farming and other agriculture related activities according to the impending weather conditions. The gain can be either through reducing input costs for farm management or through increasing.

## 2.2 Weather Forecast for Agriculture

### 2.2.1 Types of weather forecast and their applications in agriculture

Weather forecasts are mainly divided into three categories viz. (i) short range weather forecast (unto72 hours in advance), (ii) medium range weather forecast (up to 10 days in advance) and (iii) long range weather forecast (more than 10 days in advance). At present both synoptic and numerical methods are used for the prediction in the short range. But for medium range forecasting numerical weather prediction and for long range forecasting statistical methods are used.

The short and medium range weather forecasts are helpful for everyday field activities. While the medium range forecast is useful for taking measures against pest and disease attacks on crop in the field and for planning long term operational activities. The long range forecast helps in planning the future cropping system and also in arranging the required inputs. As the reliability of a forecast decreases with increasing period of its validity and decreasing area, the users should revise their plans made in advance in view of subsequent forecasts. Planning and operational activities in agriculture require forecasts in respect of the elements like expected rainfall, wind speed, wind direction, dew duration, dew intensity and variation in temperature, particularly those relating to the extreme conditions.

### 2.2.2 Weather forecasts for agriculture

Weather forecasts play significant role in decisions relating to the agricultural operation. These are normally required for sowing, planting, application of water, fertiliser and chemicals, harvesting, curing etc. and also for protection measures like those of frost, high winds etc. some of which are briefly described below.

a) **Temperature**: Growth and development of plants as well as its pests and diseases are function of temperature. Thermal regime beyond cardinal points is detrimental to the crop activity and their extremes kill the plants. Hence success of the methods to predict their growth and development hinges on the skill of the temperature forecasts. Forecasts of maximum and minimum temperature especially under heat and cold wave conditions respectively are important. Forecasts for grass minimum temperature is useful for assessing likelihood of frost conditions.

b) **Wind**: Wind forecasts are very useful for dependent agricultural operations like spray of chemicals. Forecasts of wind direction and speed should also include expected variations.

c) **Rainfall**: The probability of getting certain amount of rain is useful for planning carious agricultural operations as well as likelihood of dry weather. Rainfall predictions form the basis for irrigation management.

d) **Humidity**: Relative humidity forecast should include afternoon minima and night time maxima. Such forecasts are useful for many agricultural practices and plant protection.

e) **Dew duration**: The duration of dew occurrence is primarily used for estimating periods of leaf wetness for plant-disease control and also for determining periods for applying chemical dusts.

f) **Drying condition**: Expected drying conditions should be referred to in the forecasts. This information is used to determine crop-curing conditions. e.g., for tobacco.

g) **Sowing and planting**: Seed germination is governed by optimum soil moisture and soil temperature conditions of the surface layer (up to 10 cm) of the soil. The changes in soil moisture and soil temperature would be of interest to farmers to assess successful germination and survival of seeding. Thus forecast of rainfall and changes in soil moisture and soil temperature during the normal planting season will help the farmers to avoid sowing of seeds under such soil condition which will hinder proper germination and emergence of seeds.

**Example of soil temperature forecast:** Bright sunshine for three days will rise soil temperature sharply. Soil temperature at normal sowing depth is expected to reach and maintain level favourable for germination of groundnut and cotton seeds by early next week.

**Example of soil temperature forecast:** Light to moderate rain during the next two days will cause recharge of upper layers of soil to its field capacity. Soil moisture at normal depth of seeding is expected to reach and maintain moisture level favourable for proper wheat germination by early next week.

h) **Application of agricultural chemicals:** Application of agricultural chemicals at proper time is economical as it minimizes the losses due to release into the atmosphere. The critical factors in the effective and timely application of the chemicals are temperature, wind speed and rainfall during the next 24 hours. Wind forecast helps in determining suitable time for chemicals application. Temperature determines its effectiveness whereas rainfall immediately following application can wash off the chemicals. Agricultural chemicals which require extra attention to meteorological factors are fertilizers, nutrients, growth regulators, herbicides, insecticides and fungicides as well as chemicals used for soil fumigation and rodent control.

**Example of forecast for application of agricultural chemicals:** Wind speed mostly favourable for application of agricultural chemicals will very between 6 to 18 km/hour with varying directions becoming southerly 18 to 24 km/hour during late afternoon. If temperature exceeds 27°C, caution should be taken in applying oil based sprays.

i) **Water loss and irrigation needs**: Daily consumptive use is related to the water loss from class- A pan-evaporimeter. Estimation of evaporation loss helps in estimating loss of water occurring after the last irrigation or rainfall and hence in determining water requirements and time of irrigation for the crop in the field.

**Example of water loss forecast:** Loss of water during the last 24 hours is 0.6 cm. Expected loss of water for today and tomorrow is 0.6 cm and 0.8

cm respectively. As there is a low probability of rainfall for the remaining week, therefore supplementary water should be applied to save the crops from moisture stress.

j) **Crop curing**: The general agricultural weather forecast should provide meteorological information necessary for harvest operations and post-harvest operations such as curing. While special forecasts of certain elements are required for storage. The crucial weather factors for curing of crops are sunshine, precipitation, dew and drying rates. Precipitation may leach valuable nutrients, for example, from forage crops. Some curing processes are simply just the drying out of the crop. As for hay. Others involve complicated enzymatic processes complicated enzymatic processes that are influenced by temperature and humidity, e.g., tobacco is cured inside ventilated sheds where temperature and humidity are controlled.

   **Example of tobacco-curing forecast:** Good weather for tobacco-curing occurred during the last 24 hours. Excellent drying weather which is forecasted for today and tomorrow will cause excessively rapid drying of tobacco leaves. Sheds should be closed partly to slow down the drying process.

k) **Control of plant diseases**: Most of the plant diseases develop and spread in wet condition and the rate of development depends on temperature. Therefore, a proper vegetative wetting forecast is required for effective and economic control of diseases. This forecast includes number of hours during which crop vegetation was wet due to rain, fog or dew during the previous 24 hours. This information helps the farmers to gain maximum control with a minimum number of chemical applications.

   **Example of vegetative wetting forecast:** During the past 24 hours there have been eight Hours of wet condition with a mean temperature of 16°C. During the next 24 hours. Ten hours of wetting by rain are expected. With mean temperature of 16°C these combined wetting periods will produce a light to moderate apple- scab infection period. Growers should be prepared to commence spraying tomorrow to control the development and spread of scab.

l) **Control of noxious insects**: The life cycle of insect pests and their successive generations are largely influenced by weather factors which acting in combination. Feeding habits of insect pests are also controlled by both weather and climate. The path of insect pest migration is also largely determined by wind pattern. Weather and climatological data are very useful in determining suitable strategies. Tactics and logistics in the programs related to monitor and control of insect pests and diseases.

**Example of control of locust swarms:** Locust swarms observed near farmer in Rajasthan as soil moisture is favourable for breeding of locust. Aerial control measures are being taken and the swarms are being destroyed by ground teams.

m) **Transportation of agricultural products**: A large amount of agricultural produce is normally transported to long distance from the production place for marketing. During such transportation the temperature of most agricultural products must be kept within a narrow limit to prevent deterioration and spoilage of the products. Therefore, the heating and cooling facility of the containers must be required. The accurate forecast of maximum and minimum temperatures along the transport route is required to select proper transport equipment and its utilization.

n) **Forestry operation**: Special weather forecasts related to those weather elements which mostly affect forest management and operations such as planting of seedlings and their establishment, diseases and insects control. Forest fire, etc. are required to be issued as and when required.

**Example of weather forecast for forest fire danger:** As the forest fire danger is high today because of high winds and low fuel moisture, controlled burning and logging operation should be terminated unit rain occurs.

o) **Agricultural aviation**: Aircrafts are used for a wide variety of operations in agriculture and forestry. Because they work at low altitudes (<30m), much below of regular transport aircrafts. The specific details are not available in routine aviation forecasts. To achieve success, low-level wind drifts, stability, surface inversion (very important for ultra-low volume spray with small particle size) and vertical motions (motions more than 0.5 cm/second disperse droplets throughout the atmosphere rather than settle on the crop plants) etc. should be known.

**Example of forecast for agriculture aircraft operation:** Calm winds are expected below 90m with some patches of ground fog to 30m until 30 minutes after sunrise at 0500. Little or no wind drift is expected. The inversion is rather strong below 90m and it will break when temperatures reach 15°C at about 0800. Vertical motions will develop enough to cause ultra- low volume sprays to disperse.

p) **Protection from frost injury**: Minimum temperature forecasts are required for areas subject to freezing temperature during the crop growing season. These provide the farmer with information needed to operate his systems to prevent frost injury.

**Example of special minimum temperature bulletin:** Severe cold wave conditions are likely to affect the area. Temperature is expected to drop below 0°C.

## 2.3 Agrometeorological Advisory Services

Presently NCNRWF, generates, location specific medium range weather forecasts rainfall. Cloud cover, average wind speed, predominant wind direction, maximum and minimum temperature trends are communicated to the AAS units in real-time basis using fast communicated facilities like VSAT/ STD/FAX etc. The nodal officer in consultation with a panel of experts in various subject matters of agriculture, prepares Agro-meteorological advisories based on the medium range weather forecast provided by India Meteorological Department (IMD), Pune and disseminates these advisories to the farmers through different mass media like local newspapers, radio, whatsapp groups, personal contacts etc. and subsequently feedbacks are collected from the farmers.

### 2.3.1 Panel of experts for preparation of Agromet Advisory Bulletins (AAS)

Each AAS unit constitutes a panel of subject matter specialists from agriculture, horticulture and animal husbandry for the preparation of AAB. The panel should include agronomist, soil scientist, plant pathologist, entomologist, nematologist, sericulturist, plant breeder, horticulturist and also subject matter specialists from agricultural extension, forestry, animal husbandry etc. It is expected that experts in all the areas involve in discussion on the prevailing crop situation and anticipated weather conditions for preparation of advisories for the formers of the region.

Collection of proper knowledge on agricultural situation prevalent in the region is essential prior to the bulletin formation. Such type of knowledge includes types and growth stages of standing crops, prevailing pests and diseases, soil moisture status of the soil, state of animal health and nutrition; and prevailing agricultural marketing situation. Priority should be given to predominant crops and most prevailing problems of the region. Farm management practices like what, when and how to sow, when and how much to irrigate, what measures to be adopted for protection of crops and animals from pest and diseases, high and low temperatures, wind, rainfall etc., animal shelter, nutrition availability and their health are highly affected by the weather, hence they must find place in the AAS bulletins.

On the basis of the weather forecast received from NCMRWF and considering local agro-meteorological and crop information the subject matter specialists discuss about the consequent effects of the weather on the crops and then decide the advices for the action by the farmers. The advisory must be as simple as possible in terms of both the language and the terminology keeping in view the literacy status of local farmers.

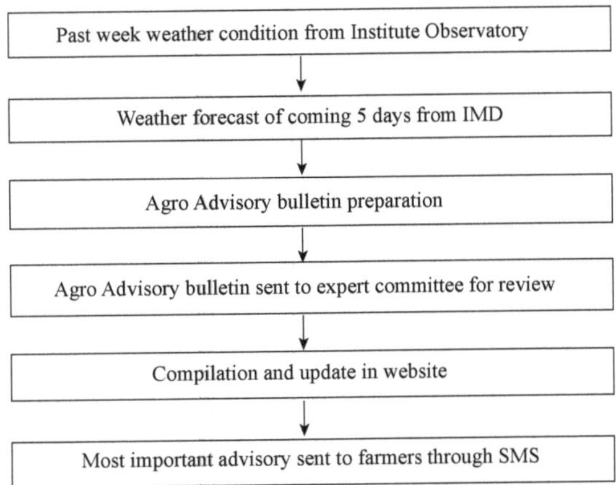


**Fig. 2.1:** Agromet Advisory Service (AAS) for Medium Range Weather Forecasting. (*Source*: ICAR- Central Coastal Agricultural Research Institute)

## 2.3.2 Dissemination of AAS

Dissemination of the weather based advisories in real time to the farmers is most crucial for success of the project. All efforts should be made in this regard to develop confidence in the advisories. Initially each unit selects about 50 progressive farmers in nearby villages and disseminates the advisories to them. Farmers' feedback will always be required for proper research work. Hence interaction with the progressive farmers will always be advantageous. As such this has to be continued. For this purpose recorded phone facility may be useful as telephone network is increasing very fast and is likely to cover even remote areas in not too distant future. However, mass dissemination of the agro advisory bulletins through AIR, Doordarshan and local newspapers in vernacular language has been found very effective and useful (Fig. 2.2). In some places, help of agricultural extension services as in case of Haryana Agricultural University and organizations associated with agricultural activities, as in case of Punjab Agricultural University, have been fruitfully utilised. AAS units should utilize services of all the organisations including non – governmental organisations and all available modes of mass dissemination at their disposal to disseminate AAS. Other than the farmers, these bulletins are also provided to the authorities of concerned departments like agriculture, horticulture, animal husbandry, irrigation, soil conservation etc. to enable them to take necessary steps for effective utilisation of the advisories.

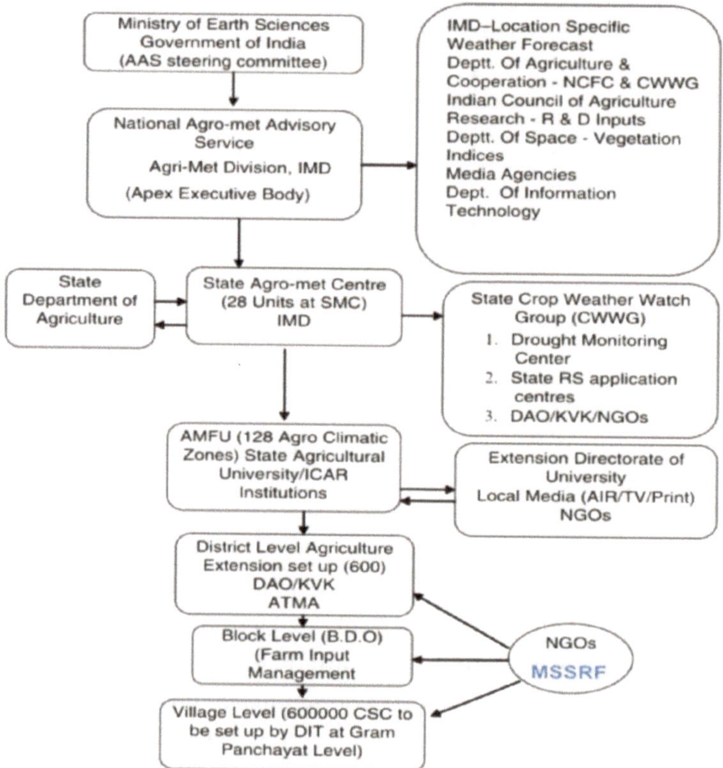


**Fig. 2.2:** Collaborating organizations and their linkages under integrated agromet services. (*Source*: IMD 2007)

### 2.3.3 Feedback analysis and verification of forecasts

In order to assess the utility of advisories and to enhance receptivity among the user community, it is essential to obtain user's feedback regularly and verify the forecasts with realized weather, and take corrective measure to improve of the forecast- advisory system. This is being done right from the beginning of the advisories under NCMRWF project. For this purpose necessary mechanism has been devolved. Farmers have also provided qualitative information regarding economic impact of the advisories in terms saving in the input costs and better production by scheduling of agricultural practices as per advisories. Table 2.1 gives the framework followed for assessing the usefulness of weather forecast through the survey and Table 2.2 describes the considered indicators of economic impact.

**Table 2.1:** Use of weather forecasts.

| Impact area | Indicator |
|---|---|
| Perception of stakeholders | Reliability, dissemination, adequacy, value-addition. |
| Awareness about AAS | Farmers knowing about AAS (%). |
| Usefulness in perception of farmers | Farmers considering AAS useful (%). |
| Use of information | Farmers following weather forecast based advisory (%). |

**Table 2.2:** Indicators of economic impact.

| Parameter | Indicator |
|---|---|
| Yield | Difference between yield of AAS and non-AAS farmers. |
| Cost | • Difference between total cost of cultivation (Rs/acre) of AAS and non-AAS farmers.<br>• Changes in cost per unit of output. |
| Profitability | Difference between return over paid out cost (Rs/acre) of AAS and non-AAS farmers. |
| Utility | Increase in utilization by farmer for manoeuvring cultural operations. |

### 2.3.4 Format for AAS

The following proforma may be followed for preparation of AAS.

a) **Weather Information:** This may include:

   i) Weather summary of the preceding week or since last bulletin including salient weather features like heavy rainfall, depression, cyclone, freezing temperature etc.,

   ii) Climatic normal for the week,

   iii) Forecasted weather information,

   iv) Drought severity index and crop moisture index for the past weeks.

b) **Crop Information:** This may include:

   i) Type of the crops,

   ii) Phonological stages of the crops,

   iii) Information about insect pests and diseases ,

   iv) Information on crop stress.

c) **Advisory:** Content of the advisory shall vary with location, weather, season, crop condition and weather sensitive agricultural practices like sowing, fertilizer application, irrigation scheduling, operations for pest and disease control etc. It should also contain special warnings for taking appropriate measures for saving crop from weather hazard, if any.

Information on location specific crop planning, crop varieties, selection of proper sowing and harvesting time etc. may also be provided. It also provides location specific suitable package and practices for cultivation of different crops.

i) Spraying conditions for weed, insect pest or disease control,

ii) Problems related to health of the animals and their products,

iii) Wildfire rating forecasts in wildfire prone areas, and

iv) Information related to livestock management like housing, health, nutrition etc. Wherever, possible outputs from crop and pest disease models may be used to increase the timeliness of irrigation applications, fertilizer applications, spraying operations etc. The advisories should provide an early warning of various severe weather phenomena such as extreme temperatures, strong winds, heavy rainfalls, floods etc.

## 2.4 Information Requirement for AAS Preparation

### 2.4.1 Non-spatial information

Readily available information on the following are essential for the preparation of AAS which may be collated and located preferably on electronic media at the AAS unit (Fig. 2.3):

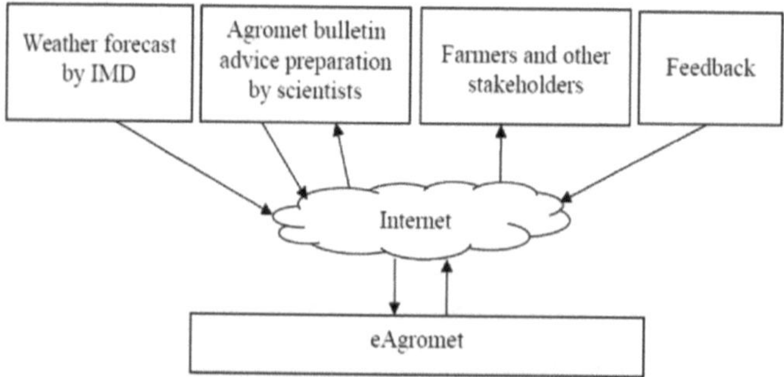


**Fig. 2.3:** AAS with eAgromet. (*Source* Reddy *et al.*, 2014)

a) Soil

   i) Soil types in the zone (base map showing major soil types)

   ii) Physical and chemical properties of the soil,

   iii) Nutrient status of the soil, and

   iv) Moisture status of the soil.

b) Crop (type, acreage and stage of growth)

   i) Level of moisture stress.

   ii) Level of thermal stress,

iii) Level of nutrient stress, and
iv) Level of pest and disease attacks.

c) Pest and disease:
   i) Intensity,
   ii) Migration trends over the region based on movements in the recent past,
   iii) Prevalent pest and disease attacks.
d) Weather:
   i) Climatic normal of the stations in the zone for the current period of AAS preparation,
   ii) Current week's weather, and
   iii) Medium range weather forecasts.
e) Weather sensitive package of practices: List of all the agro-management practices from sowing to harvests on:
   i) Field preparation,
   ii) Cultivar selection.
   iii) Seed rate, seed treatment and sowing /transplanting method,
   iv) Irrigation (time/rate/method of application), and
   v) Plant protection measure (type of pesticide, rate and method of application).

### 2.4.2 Spatial information

Base maps showing the following information:

i) Demarcated boundaries of the Agro-climatic zone, with locations of agricultural research stations, agricultural departments etc.,
ii) Major soil types,
iii) Elevation contours and slopes,
iv) Distribution of climatic parameters,
v) Major crops, land use and land cover pattern- season wise,
vi) Locations of important mass communication centres, – AAD/DD stations, new paper etc.,
vii) Irrigation network and locations of reservoirs etc.

## 2.5 Information Bank for AAS Unit

A database consisting of the following information shall be developed for ready reference for the preparation of AAS bulletins.

i) Area of the agro-climatic zone,

ii) Population,

iii) Latitudinal and longitudinal extent,

iv) Topography- maximum and minimum altitude,

v) Area covered by forests,

vi) Soil characteristics (according to USDA scheme).

# 3

# Dissemination of Agrometeorological Advisory

## 3.1 Introduction

One of the most important functions to the AAS units is to widely disseminate the advisory bulletins among the user community. A unit cannot be considered operational unless its bulletins are reaching the target masses. The dissemination is one way communication which is broadly speaking the process by which human beings share information, knowledge, experience, ideas and motivation.

Agricultural communication in India is a part of an overall agricultural extension strategy. It should not be based on trickle-down one way communication. Agricultural communication is based on the knowledge of farmer's needs. It works through existing channels of communication in a community-interpersonal. Folk-traditional and modern mass media work simultaneously at various levels such as national, regional and local. It is development- oriented and recognizes the potential of mass media and print media in achieving its objectives.

The farmers sometimes find it hard to understand and act upon complex technological innovations and as a result lag behind in adaptation to technologies. Nodal officers in association with extension agencies think of supporting system, including efficient communication for diffusing these innovations.

In spite of the fact that the rural women play an important role in farming but not much attention has been paid so far. The farm women also play a vital role in Agricultural communication. Farm women are neglected in the agricultural extension system with few exceptions. There are no special radio programmes, training classes, television programmes or demonstrations for increasing their knowledge and skills in different activities for agricultural development.

Agricultural development programmes involve the systematic application of communication technologies, methodologies and skill to disseminate agro-advisory bulletin and to receive feedback from targeted groups of farmers, and to facilitate the exchange of information among all those who are directly and indirectly involved in the programme.

There are different communication channels through which man perceives his environment. Mass media have also become very important means of communication which helps to enhance farmers understanding about weather and its forecasts. AAS units must evolve a system through which various agencies of mass communication work in the process of dissemination of advisory bulletins. The effect of three media: viz. the press, the radio and the television are more exposed hence must be utilize for the service. Answering the farmers' questions to role of mass media in rural development, what is purpose and what is the utility of providing bulleting through personal contacts to the selected group of progressive farmers.

## 3.2 Mass Media in the Rural Development

Development of rural societies mainly depends on their progress in agriculture and animal husbandry. For achieving advancement in these spheres medium range weather forecast based advisory system has been devised. All possible modes of mass communication media must be exploited to percolate advisory bulletins into the rural society.

It is found that the media have fairly high influence in creating awareness about modern practices for development. About 64.75% rural people consider media as the primary source of information about development practices. Rest do not consider media as the primary source of information about development practices.

The degree of awareness contributed by mass media is related to gender, age, education religious/caste affiliation, income, socio-economic status. Main occupation and the amount of social overheads of the area do not influence the level of adoption induced by the media persuasion.

The level of adoption is associated with education, income, socio-economic status and the nature of agricultural management practices one is using. But age, religious/ caste affiliation and social overheads of the area do not influence the level of adoption induced by the media persuasion.

The influence of education, income, socio-economic status and the nature of the main occupation is the level of adoption of development practices induced by mass media. When a fair stage of development between the developed and the less developed is achieved, the mass communication boosts it further. This indicates that the media are more useful and could be utilized for those sections of the farming community who have passed the take-off stage. The net result is the widening of the gap in knowledge and material achievement between the developed and the less developed sections of the society.

Enquiry into the association between the background variables and exposure to each of the media reveals that there is no difference in the extent of exposure to the press on account of the difference in age. But regarding the radio, the older section of the people is more exposed to the media. The younger group is more exposed the television. Religious/caste affiliation influences the people's exposure to all the media. The exposure to each of the media is found to increase with increase in educational attainment. The income also augments the farmers exposed to each of the media. The radio and television increased with increase in socio-economic status. The nature of main occupation is a determining factor of the people's absorption of the communication to the individual. No association was found to exist between exposure to any one of the media and the amount of social overheads of the place of residence.

## 3.3 Exposure of the Rural People

Every medium has its own specialities. The illiterates are not exposed to the press. Even if exposed it is through the literate. The television requires its audience to be collected together at one place at fixed times. Moreover it's been discharging mainly the recreation function. The radio does not have many of these drawbacks. However, a major shortcoming of the radio is that the listener cannot listen to the radio broadcast according to his convenience. Irrespective of the literacy level to the people, topography and geographical location of the area of residence, the radio reaches all rural people easily the recreational value of the radio along with its affordable price and portability increases the attraction of rural people to this medium.

### 3.3.1 Time spent daily for reading newspaper

It is seen that many rural people do not read newspapers for one reason or another. Illiteracy was found to be the main reason for not reading newspaper. Financial constraints prevent certain literate respondents from reading newspaper interestingly. It is seen that few illiterates are exposed to the newspapers through others. They expose themselves to newspapers when they visit their local tea shops for morning tea where the papers are read loudly by someone. Those who are in the habit of reading newspapers daily spend on an average 1.25 hours for this purpose. The standard deviation of reading time is 0.52 hours. These facts show that there is substantial difference in the time spent by the respondents for reading newspapers.

### 3.3.2 Time spent for listening to radio programme

Analysis shows that the rural people are more exposed to the radio. The major reason for this is the absence to the literacy barrier for radio listening.

### 3.3.3 The Television and rural people

The television is a potential medium for propagating agro-meteorological advisories. It also inculcates new the impact of television on them. Farmers watch television not merely for entertainment but it is viewed as important source of information on weather forecast and farm management. This medium attracts people belonging to all categories.

## 3.4 Modes of Bulletin Dissemination

Since the inception of the service, AAS units have been periodically obtaining feedback on the issue of media through which farmers want these bulletins. The feedback analysis suggests that they are more exposed to three media: viz. the press, the radio and the television. Therefore, ideally all the three media should be utilized for reaching the largest possible number of the farmers. Among the three medium, radio is the most preferred because of portability, cost, network etc. Regularity and frequency of the bulletin dissemination are two key issues. As far as print media is concerned, farmers would like to have a special column/ section of the local newspaper bulletin. Hence, efforts may be made by Nodal officers to get Agromet Advisory Bulletin (AAB) at a pace on every Tuesday/ Wednesday and Friday/ Saturday. Unfortunately, most of the local newspapers are printed during night time hence, the bulletin is published with a time delay of 24 hrs. It can be reduced significantly through broadcast of AAB on All India Radio (AIR). All the AIR stations have a slot for farmer's programmes during different times of the day. As AIR has got a fairly dense network is comparatively sparse and one station caters to entire state or even more at some places. There is need to composite the AAB for the entire area under the jurisdiction of the concerned of the concerned state. It can be effectively done by the SAU Headquarters or the AAS unit located near the Doordarshan station. To enable it, all the units should fax their bulletins to the designated officer in time.

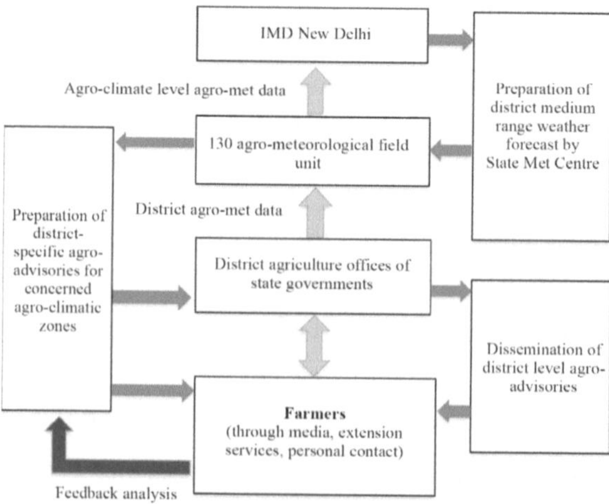


**Fig. 3.1:** AAS institutional mechanism to reach farmers. (*Source*: IMD Progress and assessment of AAS program Progress and future plans)

## 3.5 Dissemination through Personal Contacts

This mode of communication is mainly for getting first hand feedback on the worthiness and usefulness of the service. It also helps in assessing the economic impacts of AAS. It can be achieved through following modes (Fig. 3.2).

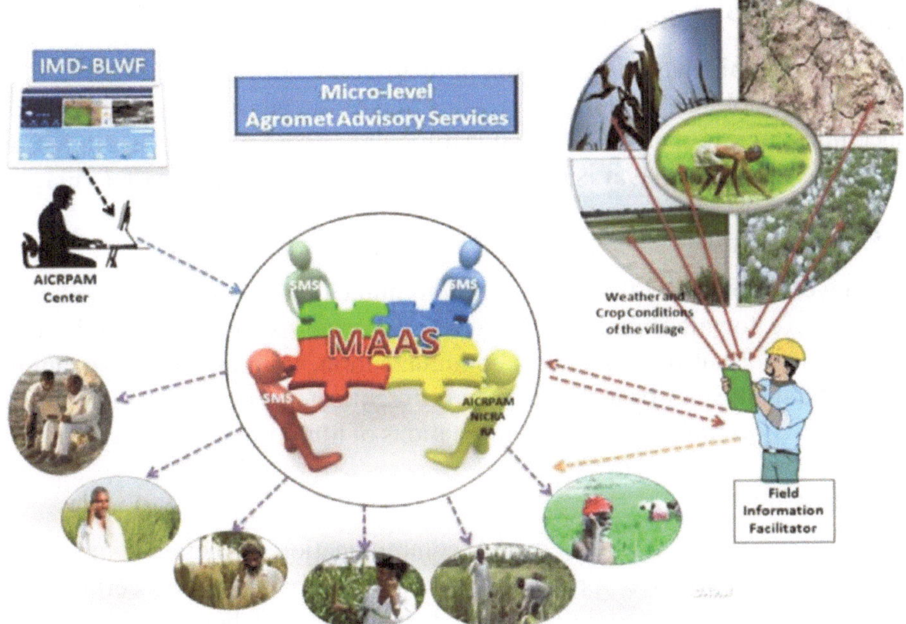


**Fig. 3.2:** Development and dissemination of Micro-level Agromet Advisory Services (AAS). (*Source*: Vijaya Kumar *et al.*, 2017)

### 3.5.1 Courier / Postal

Agromet Advisory Bulletin can be sent through courier on a specified bus route or a hired vehicle. During the process of delivering the bulletin, the technical officer may also collect information on the prevailing agricultural situation (including status on pest disease and soil moisture stresses) and what action the farmers took on the provisory.

### 3.5.2 Farmer's fair

Farmer's fair is other element of the communication support programme that serves as bridge between agricultural experts and farmers. It delivers more information and new technology to rural people. In farmers fair, exposure to the educational posters is most visible enabling them to get access to the information. Through which utility of AAB can easily be communicated.

### 3.5.3 Telephonic contacts

Telephone conversations are similar to face-to-face conversations in many regards. The telephone is more ecologically valid for dissemination of AAB, even farmers can easily ask for AAB. AAB can be disseminated by alternative tools like wireless etc.

### 3.5.4 Gram Panchayat

Local leadership has a paramount role in rural development and it is an appropriate political / administrative institution at the local level. The local specificity is important otherwise it is difficult to adopt AAB by small farming families. Gram panchayat is most effective link with the people of a village.

### 3.5.5 Media mix

It is a well-known fact that, no single communication source or channel could be effective for all situations. Suitable combinations of media (media mix) become important in agricultural communication. Its merit lies in its ability. agricultural advisory or technology to different categories of farmers. Overall effective scores ranked high for combinations of film+ Folder + Demonstration.

## 3.6 Future Plan

It has been found that the media communication is better promoter of Agro Advisory Service for those farmers who have attained certain level of advancement in adopting recent technological advancement in farm management. The less advanced categories are benefiting less by development and communication. Their absorption of the information can be increased only if they are brought to the take off stage on the socio-economic front. For this the Nodal Officers in association with extension agencies have to concentrate their attention on these sections to the society.

## 3.7 Credibility

Practicability and timeliness of AAB are important criteria for acting upon them. Hence, the massages should be clearly worded, be adequately descriptive and trustworthy compensated to induce innovativeness and experimentation. Imaginative thinking is required for devising a suitable set up for implementing this in practice. This will gradually augment the credibility of the media among the general public. AAS system is required to save the time loss at different steps.

It is seen that the correlation between mass media exposure and development is the highest for the agricultural. This indicates that the mass media are main source of transformation. This is especially true in the case of radio. It is seen that the rural people are more exposed to this medium. Moreover, it is revealed that there is no relationship between education and exposure to the medium. Hence, it is most suitable for spreading information among rural people who are lesser educated. In view of this fact the quality and coverage of broadcast meant for rural people are to be avoided in the broadcasts. Formation to very sophisticated language is to be avoided in the broadcasts. Formation of rural forums for conducting development broadcasts may help to alleviate the difficulties created by the lack of personal contacts between the communicator and the receiver.

To ensure success of this project, it will be essential to ensure that the AAB disseminated is demand and user driven. Information on the entitlements of rural families in areas affecting their day to day life should be made available.

# 4

# Weather Forecast and Livestock/Poultry Management

## 4.1 Introduction

The livestock/poultry have great capability for climatic adaptation and acclimatization. However in an open environmental condition their overall well-being and efficiency depends on day-to- day weather conditions. While certain breeds of livestock and poultry are considerably flexible to adjust themselves with the varying weather conditions, the others may have little adjustability. This has led to the development of weather/climate based livestock/ poultry management practices which include the general care. Nutrition and health managements for optimizing conditions for best yield in respect of milk, meat, eggs, wool, load carrying capacity etc. have been properly documented. The farmers can be advised about appropriate measures to be taken by them to minimize the effects due to advised about appropriate measures to be taken by them to minimize the effects due to adverse weather conditions on the basis of weather forecast. But the places where the livestock and poultry management practices have not been properly documented, this is to be done for the success of advisory related to the livestock/ poultry management.

## 4.2 Requirenents for the Development of Livestock/ Poultry Management Practices

Following studies are carried out for the development of local livestock/ poultry management practices.

### 4.2.1 Identification of appropriate breeds

In order to improve their gains incur losses interested in newer breeds. However, in many cases they adopt newly opted breeds with the local environmental conditions. Therefore, before introducing a new breed in certain area, optimum weather conditions for those are to be decided detailing the resistance factors and the range of various meteorological parameters like temperature, humidity etc. in which they can adjust.

### 4.2.2 Nutrition and feeding considerations

The feed intake is essentially dictated by the range of ambient temperatures in which a normal body temperature (zone of thermo-neutrality). The essentially constant lower limits of temperature in which acclimatization is possible depends on age, sex, species, health and body condition, nutrition level, time of the day etc. The temperature forecasts can be paramount important in nutritional management.

Normally, feed consumption is higher during cooler periods than the hotter periods. When the temperatures are high sometimes they stop eating for several hours. In such a situation they may over eat when the temperature drops and this may lead to risk for health. Proper feed intake may be maintained by feeding them during cooler times of the day. Feed must be excess or poor quality roughage, fat, protein, mineral etc. Usually feeding of excess or poor quality roughage increases heat production. Minimum levels of roughage must be maintained to support active rumen function and minimize acidosis. This layer is affected by roughage type, grain type and form the inclusion of ionospheres and possibly buffers and the feeding reduced by fat in dry rations making them more attractive. The protein intake must be watched carefully because underfeeding decreases intake, growth and production; white overfeeding increases heat production and stress. Protein metabolism and the elimination of excess protein and urea are exothermic. Therefore, in hot weather, heat production can be lowered by increasing protein quality while reducing the total content. Similarly, Protein supplements with lower solubility or rumen degradability can be advantageous as compared to highly degradable protein. Increased water loss and evaporative cooling through panting will decrease the blood carbon dioxide levels and increase the blood pH, thereby causing respiratory alkalosis. Adding acids or salts (e.g. HCI, KCL, K, SO) i.e. adding ions has as an acidogenic effect. It alters the digestive acid-base and mineral status and increases the water intake substantially thereby reducing heat stress in cattle. This is especially occurred when the water temperature is less than the body temperature. It is also similar in broilers, but the supplementation must be monitored carefully. When it is hot, cattle normally prefer moist or wet feeds, such as silage. But wet rations can quickly dry out and become stale and unappetizing in hot weather. The situation is exacerbated when cattle are fed in a pattern encouraging wastage, i.e., overfeeding or feeding increases the dry matter intake by heat-stressed poultry. The feed management problems presented and compared with those of dry feed and may outweigh any advantages. The contents of the feed should be suitably adjusted in rhythm with the prevailing and predicated weather conditions to keep the livestock/ poultry in best state from the point of view of expected yield them.

Cold weather increases the maintenance requirements of animals for acclimatization. In some areas where quality roughage is in short supply or grain is cheap, cattle may be fed with only 7-9% roughage. Roughage intake need to be suitably increased during winter to maintain a consistent intake and to reduce the risk of acidosis. Breeds and hybrids with poor cold tolerance are still acclimatizing to a colder environment benefit from a higher roughage diet, say 20-25%, because of more heat production. For some rations, adding about 3% molasses may temporarily raise feed intake.

To maintain feeding levels, drier feeds need to be offered as they provide better dry matter intake level, for example, silage with 60-65% moisture compared to moisture of >67%, similarly grains with 24-26% moisture instead of 28-30%. Alternatively, dry feed such as hay and dry grain can be included in the ration. For a short term effect on appetite in levels can be decreased say from 28g to 25g per ton. The feed troughs should be regularly checked to make sure that uneaten feed is not building up. These "fines" are less palatable and will get used to what they are offered, fresh, appealing and appetizing feed makes for a less complicated and more stable production than musty, mouldy and unpalatable feed.

### 4.2.3 Health of the animals

Advisory must include advice on livestock medications and as far as possible a veterinarian must be included in the advisory board. The Technical officer should include advisory on animal health status and control measures. Medications and antibiotic can be included in livestock and poultry feeds. With a rise in temperature and decreased in feed intake there will be a concomitant decrease, for example in the intake of anticoccidial drugs by poultry. Raised temperatures also negatively affect the development and sometimes make some toxic products, for example inclusion of nicarbazin at the normal dose of 125 ppm will produce excess poultry deaths following heat stress. This was ably demonstrated in Georgia during the summer of 1980 when the ambient temperature exceeded 38°C for eleven days: broilers on nicarbazin suffered a 91% heat-stress mortality compared with only 27% in non-medicated birds. Mechanism nicarbazin exaggerates the body temperature response to heat stress. Amprolium a widely used anticoccidial, is a specific thiamine antagonist. Defective mixing or the inclusion of merely three times of the normal level can produce vitamin deficiency. The risk is exacerbated by an increased need for thiamine at raised temperature. In the absence of proper feed production, storage and antimetabolites, marginal deficiency may result without the synergistic deleterious effects of amprolium. Thiamine deficiency also produces anorexia and thus further problems.

### 4.2.4 General care

Cattle can adapt to sustained temperature changes. But adaptation must not be rushed. Broilers allowed to adjust to high diurnal temperatures will maintain a body temperature approximately 10°C lower than non-acclimatized birds and suffer little mortality. In order to avoid under stress, environmental conditions have to be artificially adjusted and necessary protection is to be provided from heat, cold, wind, mosquitoes / flies / insects etc.

Anything that will ameliorate temperature stress can benefit livestock. Shades and shade trees have proved advantageous in some areas, especially where there is high humidity, intense sunlight and little wind. Cooling and sprinkling water to provide evaporative cooling will alleviate heat stress and improve performance and intake. With sows evaporative cooling is a more effective technique for relieving heat stress than snout coolers, stotted floors or high energy-density diets. Cooling during the dry period before calving can increase milk production in the following lactation. It is traditional in some areas and countries to provide pools for cattle and pigs to lie in, while pools may cool stock. They can also provide an ideal breeding medium for disease vectors (e.g. mosquitoes and snails) and for culturing water- borne zoonotic infections. High humidity, light or calm wind and warmer temperatures can provide a hazard by attracting mosquitoes and flies unless cattle are sited to take every advantage of breeze and wind. Forced ventilation has important impacts on maintaining intake. Cool breezes also reduce the insects. However, a breezy, windy country which can be very productive in hot months can be very uncomfortable during winter. If cattle are kept near their pens and grazing, they will become excited and their heat load will be increased. They will waste energy battling this continual nuisance and spend less time feeding. Site management and the seasonal use of chemically treated ear-tags are profitable.

Cattle with long hair coats or which have not shed hair will be less heat tolerant and demonstrate lower conception rates in the spring and summer breeding seasons. This can be associated with grazing endophyte- infested fescue, heavy parasite infestations and the long term effects of foot and mouth disease. Hence advice on appropriate time for cutting the hair in case of sheep, goats, camels etc. may find place in the bulletins.

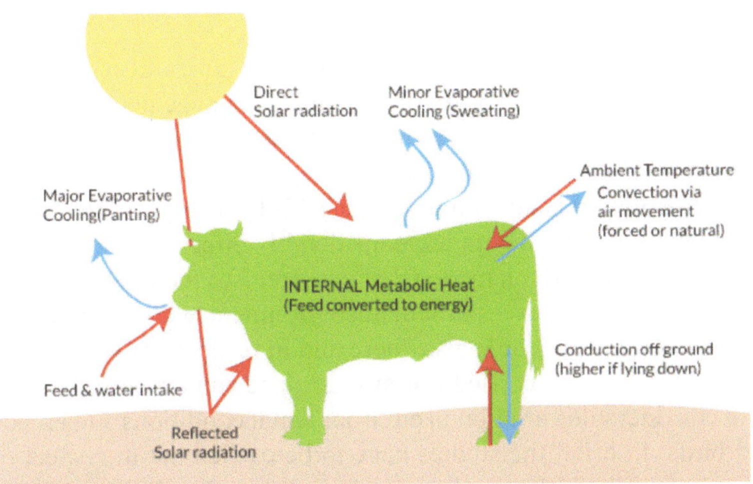


**Fig. 4.1:** Energy balance of an animal.

### 4.2.5 Clean, fresh and cool water

Ready access to cool water from automatic water resources and the water from stock tanks or ponds at ground temperature is much preferred to sun-feed intake water. As water intake declines, feed intake and water intake tend to be a function of dry matter consumption and ambient temperature. Water serves as a heat buffer that reduces a potential rise in body temperature. In poultry water losses increase via respiration as the temperature rises. But if cool (13°C) water was provided, there was a 10°C drop in body temperature as well as simultaneous local overcrowding. Poultry can tolerate higher temperatures and maintain a normal 41°C body temperature when more water is provided. It should be noted that overcrowded birds have difficulties in maintaining their body temperature even at high intake of water at house temperature of 15°C to 20°C.

### 4.2.6 Marketing

Finished cattle can be very susceptible to the effects of hot weather both in the final weeks of feeding and in shipping. Marketing stock with less finish and at lighter weights may be financially more advantageous and less costly during high temperature. In many areas, movement warnings are now issued to hog shippers when the weather is dangerously hot. Magnesium aspartate hydrochloride has been successfully used to reduce stress and losses when transporting swine, though it's potential for poultry is unclear. It seems that exposing broiler cockerels to a 24 hours heat stress of 35°C at five days old may significantly reduce heat stress later in life and especially during marketing.

## 4.3 Role of Agrometeorologists

The agro-meteorologist has to study the effect of variation in weather conditions on the animals and birds in order to evolve suitable management practices for optimising the environmental conditions and taking suitable measures to minimise the ill effect due to weather. These studies will help in evolving the mechanism for advisories related to animals and birds. The performance of animals, their productivity and health are considerably effected by variations in weather. The adjustments of tolerance from one breed to the other, thus the results of a study undertaken in a particular environment with reference to certain breeds may not be applicable elsewhere and in respect of every breed. As such, the results of studies carried out at one place cannot be utilized at other place with considerable variation in environmental conditions and breeds of animals and birds. In brief, the studies have to be carried out in respect of breed difference, colour, body formation, metabolism, genetic difference, age, productivity, physiological studies and nutrition parasite, disease and other stresses. Some of the topics for the studies may be as under.

a) Optimization of environmental conditions for different animals including housing aspects.

b) Limitations with respect to variation in different meteorological elements.

c) Conditions in which diseases are likely and productivity of animals get adversely affected.

d) Studies relating to gestation, age and body conditions like lactating or dry condition and the performance of load carrying animals.

e) Heat budget of animals, and

f) Feeding and hygiene conditions

Through cycles of acclimatization with change in seasons, animals get adjusted with a given environment but the adjustment limit varies for different breeds and different body conditions. Thus humidity, thermal radiation, air velocity, light, precipitation, pressure, dust and gases are to be taken into account to determine the suitable environmental conditions. The environmental criteria development requires quantification of environment, performance measures and biological exposure function considering threshold limits, effects of compensation and statistical evaluation of environmental factors. An agro-meteorologist has to collect data not only about the environmental conditions but also about the animals and birds. The studies will help in evolving mechanism for advising the farmers to take effective measures.

Because of wide individual variations, the statistical descriptions representative to the population should be used. The historic measures of growth, milk, eggs, wool, reproduction, feed conversion and mortality serve as parameters for the economic assessment. The animals' performance should be judged from an optimum level dictated by the local conditions. As the management improve, the productivity approaches the potential level. Since the nutritional level and feed intake depend on environmental conditions the upper and lower critical temperatures have to be worked out accordingly. Within threshold limits the animals have considerable ability to rebound when stresses are removed. As such the adjustment with respect to feed should be decided.

The aim of forecasting animal diseases is to provide the farmers an opportunity to take effective measures to avoid losses due to animal diseases. As the diseases are related to environmental conditions the accuracy of forecasting is necessary for disease forecasting. Some of the important advantages of disease forecasting/ advisories are:

### 4.3.1 Strategic

The planning and direction of health programmes are aided by seasonal forecasts such as for fascioliasis and nematodirasis and for airborne infections such as foot-and-mouth disease.

### 4.3.2 Tactical

Knowing the dates for maximum larval hatch and tick rise can optimise the use of helminthicides and acaricides reducing the volume of unnecessary dose and the cost.

### 4.3.3 Additive

In areas with chronic problems, such as mineral deficiencies, prophylactic measures, warnings to the provision of minerals and other conditions such as swayback of increased numbers of outbreeds due to congenital copper deficiency add to and rein being given by others.

### 4.3.4 Supportive

Specific warnings and advice on preventive care will reduce helplessness and increase agricultural/ meteorological contacts.

### 4.3.5 Research

The research effort to produce forecasts or implement existing system will clarify disease cycles and identify specific questions for which answers are needed; this is especially fruitful when analysing why forecasts were not correct. Because the research frequently uses existing data, its costs are minimal while the benefits, if properly utilised, are very large.

### 4.3.6 Data base

While meteorological data are usually available in adequate detail, regular reports of animal diseases are sparse and frequently biased. The development and use of disease monitoring both in the provision of improved veterinary care as well as in epidemiological control and knowledge of disease are required.

# 5

# Agromet Advisory To Combat the Effect of Extreme Weather Events

## 5.1 Introduction

Weather is one of the most important factors which determine success or failure of agricultural production. It effects every growth phases and development of plant. During the growing period of the crop, any variability in the weather condition such as delay in monsoon, excessive rainfall, drought, flood, heat or cold wave affects the crop growth and finally reduces the quality and quantity of crop yield. These losses due to weather hazards can be reduced by taking proper crop management in time and by applying accurate weather forecast. Weather forecast also provides guidance for selection of crops and varieties best suited to the local climatic condition. The principal objective of the weather forecast is to inform the farmers about the actual and expected weather condition and its impact on the various day-to-day farming operations i.e. sowing, irrigation scheduling, weeding, time of pesticides spray and fertilizer application etc. Weather forecast also helps to increase agricultural output by reducing losses and risks. It helps to increase efficiency in the water use, labour and energy, reduce costs of inputs, improve quality of yield and also helps to reduce pollution with judicious use of agricultural chemicals. Rathore *et al.* (2001) discussed the weather forecasting scheme operational at National Centre for Medium Range Weather Forecast for issuing location specific weather forecast for five days in advance. Damrath *et al.* (2001) mentioned that statistical interpretation methods are used to increase the reliability of the forecast on precipitation. One of the key challenges in the 21$^{st}$ century is that Agromet Advisory Services (AAS) have to respond to an unpredictable and changing climate. In India Gramin Krishi Mausam Sewa (GKMS) is the umbrella project for Agromet Advisory Services (AAS) and it intends to link AAS to different sources of knowledge and innovations to respond to the climate change at root level farmers. The main aim of GKMS is to create effective, efficient and synergistic linkages to improve the delivery of these AAS to the farmers. GKMS works under the frame work of India Meteorological Department (IMD) which seeks to enhance the livelihood of every Indian farmer. It directly focuses on the needs of the farmers, contributing to sustainable growth in and transformation of Indian agriculture by providing weather based effective advisory services. It introduces the opportunities and challenges for AAS in response to changing climate and begins to outline the possible roles of adaptive AAS.

Continuous increasing of uncertainties in weather and climate acts as a major threat to the food security of the India. Besides, the possible impacts of climate change also create major challenges in agriculture sector in the country. The combination of long term changes in weather condition and the higher frequency of extreme weather events also have adverse impacts on the food production in the coming decades. India Meteorological Department has taken major initiative to implement state of art technologies which are essential to identify weather and climate related issues on Indian agriculture and also to realize the present condition, needs and demands of the poorer farmers of the country. India Meteorological Department (IMD) has started India Agro-meteorological Advisory Service (IAAS) in the country for the benefits of farmers. AAS is an innovative step by IMD, Ministry of Earth Sciences to enhance crop production by providing real time crop and location specific agromet services on crops and livestock management strategies and operations to village level. It has a potential to change the face of India in terms of poverty and food security.

The AAS provides special kind advisory bulletins for the farmers. It has made a huge difference to the agricultural production by taking the advantages of favourable weather and minimizing the adverse impacts of extreme weather events. IMD launched the scheme Integrated Agromet Advisory Services (IAAS) in collaboration with different institutes/organisations/ stakeholders from 1st April, 2007 for weather based farm management. Under IAAS, a mechanism was developed to integrate weather forecast and agro-meteorological information to prepare weather based agro-advisories which significantly help to improve the food security in India by enhancing farm production and productivity.

## 5.2 Structure of Agromet Advisory

The Gramin Mausam Sewa (GKMS) project is implemented through five tier structure to set up different components of the service spectrum. It includes meteorological (present and forecast weather), agricultural (identifying weather sensitive stages and preparing suitable advisories based on weather forecast), extension (two way communication with farmers) and information dissemination agencies (telecom, media, information technology).

## 5.3 Weather Forecast Based on Seven Parameters

Quantitative district level weather forecast up to five days was started rom 1st June, 2008. The quantitative forecasts are given for seven weather parameters viz., rainfall, maximum temperature, minimum temperature, wind speed, wind direction, relative humidity and cloud cover. In addition, weekly cumulative rainfall forecast is also provided. IMD, New Delhi generates these weather forecasts using Multi Model Ensemble technique based on different models available in India and other countries. These weather forecasts are disseminated to Regional Meteorological Centres and Meteorological Centres of IMD located in different states of India. After value addition using synoptic interpretation of model output, these weather forecasts are sent to 130 Agro Met Field Units (AMFUs) co-located at different State Agriculture Universities (SAUs), institutes of Indian Council of Agriculture Research (ICAR), Indian Institute of Technology (IITs) etc., for preparation of district level agromet advisories on every Tuesday and Friday.

## 5.4 AAS Bulletins at Different Level

The Agromet Advisory Bulletins are issued at block, district, state and national levels. The block and district level crop specific advisories including field crops, horticultural crops and livestock bulletins are issued by AMFUs. The State Level bulletin is jointly prepared by State Meteorological Centre of IMD and AMFUs; and it is a composite of district bulletins helping to identify the distressed districts of the state as well as to plan the supply of appropriate farm inputs such as seed, irrigation water, fertilizer, pesticides etc. Ministry of Agriculture is the prime user of these agro advisory bulletins. Important decisions are taken in weekly Crop Weather Watch Group meetings (CWWG) conducted by Ministry of Agriculture at national level and ASS bulletins provide significant inputs to this CWWG meeting. Bulletins are also used by irrigation department state government and also used by a large number of agencies including seed corporation, fertilizer industry, pesticide industry, transport and other organizations which provide inputs in agriculture. National Agromet Advisory Bulletins are prepared by National Agromet Advisory Service Centre, Division of Agriculture Meteorology, IMD, Pune, using inputs from various states. These bulletins help to identify stress on various crops for different regions of the country.

District level medium range weather forecast and advisories help to minimize output loss and crop damage. It also helps the farmers to plan for irrigation scheduling, weedicide and pesticide applications and many more weather related agricultural operations including selection of cultivar, dates for sowing or planting, dates for intercultural operations, dates for harvesting and also post-harvest operations. Agromet advisories help to increase profits by consistently delivering actionable weather information, analysis and decision support for farming situations such as:

- Diseases and insect pests management through forecast of temperature, relative humidity and wind speed,
- Irrigation management and scheduling through rainfall and temperature forecasts,
- Protection of crops from thermal stress through extreme temperature forecast.

An ideal Agromet Advisory Bulletin enables the farmers to reap benefits of benevolent weather and to minimize or mitigate the impacts of adverse weather.

- District level weather forecast for next five days for different weather parameters like rainfall, maximum and minimum temperature, wind speed and direction, relative humidity and cloud cover, including forewarning of hazardous weather events (drought, flood, cyclone, hailstorm, heat/cold wave etc.) likely to affect standing crops and suggestions to protect the crops from these weather hazards.
- Information on soil moisture status and appropriate guidance for application of irrigation, fertilizer and pesticides etc. based on weather forecast.
- Advisories on suitable dates for sowing/ planting of crops and carrying out intercultural operations covering the entire period from pre-sowing to post harvest to guide farmers in their day–to-day field operations.
- Weather forecast based forewarning for incidence of major insect pests and diseases of crops and advises on plant protection measures.
- Different management practices like shading, mulching, shelter belt, frost protection etc. for manipulation of crop's microclimate to protect crops under stressed conditions.
- Reducing agricultural contribution to global warming and environmental degradation by judicious management of land, water and farm inputs particularly fertilizers, pesticides and herbicides.
- Advisories on health, shelter and nutrition management for livestock.

On the support of above mentioned topics, district specific Agro-meteorological advisory bulletins are prepared to meet the farmers' need and to support their decision making processes. The suggested agro advisories generally alter farm activities in a way that improves outcomes as it contains advice on farm management practices aiming to take advantage of favourable weather condition and mitigate the stress on crops and livestock. The bulletins are made in a format and language which is easily understood by the farmers. The agro meteorologists first interpret past weather and the forecast weather for next five days and then translate it into layman's terms so that farmers can understand it easily. For making the advisory bulletins, agro meteorologists use state-of-art technologies such as different crop weather models, climatic risk management tools, GIS generated agromet products etc. Interaction between AMFU members and farmers is also promoted through participatory approach to identify the weather sensitive decisions. This step develops a relationship between the IMD, AMFU and farmers which helps to identify or diagnose the gaps in weather information and services available from the IMD.

ASS includes a number of climate resilient practices such as small water infrastructure, watershed management, land use planning, intercropping, agroforestry, low tillage and soil erosion. The agricultural policies are mainly based on the premise of increasing agricultural productivity to reduce poverty by increasing economic growth. This is being implemented alongside major statements regarding food security. Current existing policies are generally supportive of agricultural practices (example: expansion of cultivated land, increasing farm mechanization, use of organic fertilizer and other inputs) that focus on increasing short term production. But they are less supportive of agricultural practices which improve food production, enhance adaptive capacity and address mitigation (example: restoration of degraded land, improving soil micro and macro nutrients).

## 5.5 Characteristics of AAS and Exploration of 'Adaptive' Attributes

Over many decades in India, regular use of old ICTs (such as telephone or television) was rare for large sections of the rural populations. The most striking exception was radio, which quickly became widespread due to the availability of cheap battery-powered and transistorized receivers. GKMS have maintained communication departments and produced regular radio programmes on agricultural topics for broadcasting to rural populations, often through state-owned radio stations. In some cases, educational videos or TV programmes were also produced for screening either via mobile audio-visual vans or on the often state-run TV stations. Content was largely created and controlled by the AAS organizations and targeted at the farmer recipient. Since the turn of the millennium, meteoric growth of private mobile-phone ownership and use in both rural and urban settings, increasing access to TV and video-screening facilities, and digital filming apparatus (cameras, mobile phones), and the more recent spread of internet access in towns and even into smaller towns and canters via mobile net services, have offered a whole new world of opportunity for multi-directional communication. Interestingly, in our experience, even while AAS has embraced newer participatory approaches such as farmer field schools and farmer participatory research to mobilize communities and harness complementary contributions from researchers, farmers and AAS staff for innovation. AAS in general seems to have been relatively slow to explore opportunities for a comparable revolution in multi-stakeholder information sharing, knowledge creation and advocacy activities offered by combinations of new and old ICTs. A major challenge for ICTs in AAS vis-à-vis climate change issues will be to creatively develop ICTs as multi-way platforms and break with the unidirectional communication traditions of the past. It is not only AAS staff who are in need of information and perspectives about climate change science and their expression in their local environment. Researchers and official meteorological stations are one source of these, but both AAS leaders and researchers also need to access and learn from the experience of farmers, frontline AAS staff and other sector staff living and working in the focus areas. Adaptive capacity varies widely between individuals and communities, due to differing access, control of assets and the institutional environment in which people are living. In order to strengthen adaptive capacity, GKMS need to be able to recognize these differences and develop strategies to address them. Self-organization is a key element of adaptive capacity. Climate change has emerged only recently as a critical issue and so most AAS individuals would have received little specific training in relation to climate change in their formal training. This is starting to change, but many AAS sectors have limited capacity to actively seek and use new knowledge and information. This is a critical factor that will have a major influence on the extent to which AAS will access information and networking initiatives.

## 5.6 Agromet Advisory Methods

In moving towards adaptive AAS, the used advisory methods are critical. In dealing with climate change and other uncertainty, emphasize should be given on some aspects such as: strengthening the capacity of clients/farmers rather than delivering messages, enabling the farmers to use weather information in their day–to-day field works, strengthening the self-organization of farmers, improving links between researchers and extension workers, enhancing local level innovation and considering the content of advice in relation to what is appropriate to the context (ex, balancing production-innovation, growth and climate resilience).

## 5.7 Gramin Krishi Mausam Sewa (GKMS)

The agromet advisory service based on medium range weather forecast has been made operational by National Centre for Medium Range Weather Forecast (NCMRWF). NCMRWF has proposed to establish 130 Agro Meteorological Field Units (AMFUs) in all the Agro climatic zones, which cover the entire country. The agro advisory bulletins are prepared by the panel of scientists based on the "Package of practices recommendations developed by the University for the Respective Zones. The essence of agro advisory is to make the farmers aware of farm operations for sustainable agricultural production based on weather forecast. District wise weather forecast for next 5 days is received twice in a week on Tuesday and Friday. Based on total forecasts received a no. of Agromet Advisory Bulletins were prepared and circulated among selected farmers of different villages of the region. Weather forecasts and agro- advisories were also communicated telephonically to some of the nearby farmers of all the districts of Bihar by different AMFUs working under GKMS project. Even District Agrometeorological Unit (DAMUs) existing in KVKs are sending the agro advisory to the farmers and different stakeholders (Fig. 5.1) for immediate benefit of the local farmers. For e.g. All India Radio, Bhagalpur broadcasts Agromet Advisory Bulletins in between 18.30 and 19.00 hours on every Tuesday and Friday of the week. Similarly different Newspapers like "Hindustan", "DanikJagaran", "DainikBhaskar", "PrabhatKhabar", "Rashtriyasahara", and "Nayibaat" published weather forecasts and Agro-advisories on regular basis. Weather forecasts and agro-advisories are also sent by email/ faxed to Annadata Programme of ETV, Sahara TV, Sadhna TV for telecast, for the benefit of farmers of the state. Agro advisories are also sent by email to IMD, Pune and uploaded on website of the Agromet Division of IMD and Website of different universities existing in Bihar and likewise in other states of India. Agromet Advisory Bulletins are sent to Joint Director of Agriculture and all District Agriculture Officers of different districts of state. zone III A& III B, Project Directors of ATMAs of state, Block Agriculture Officers, for communication among the farmers through Village Extension workers or local medias. The college also organized Kisan Mela, Kisan Gosthi and Farmers Awareness Programme in which the visiting farmers were made aware about this service. Weather forecast and agro advisories are also sent through SMS on mobiles of Farmers on regular basis. For e.g. in Bihar, a no. of SMSs had been sent and more than ten million farmers had been benefited and more than 15 lakh farmers from 38 Districts of Bihar has been registered and are getting SMS at a time.

**Fig. 5.1:** Agro Climatic zones of Bihar. (*Source*: Indrani Dana *et al.,* 2013)

**Table 5.1:** Agro climatic zones of Bihar and AMFUs of GKMS.

| S. No. | Agro climatic Zone | Districts | Agromet Field Unit (AMFU) |
|---|---|---|---|
| 1 | North West Alluvial Plain Zone (Zone-I) | West Champaran, East Champaran, Gopalganj, Vaishali, Sitamarhi, Muzaffarpur, Siwan, Darbhanga, Samastipur, Sheohar, Begusarai, Madhubani, Saran. | Pusa (Samastipur), DRPCAU, Pusa. |
| 2 | North East Alluvial Zone (Zone-II) | Saharsha, Purina, Katihar, Supaul, Khagaria, Madhepura, Kishanganj, Araria. | Agwanpur (Saharsha), BAU Sabour. |
| 3 | South Bihar Alluvial Zone (Zone III A & B) | IIIA: Bhagalpur, Lakhisarai, Munger,<br>IIIB: Bhojpur, Buxar Nalanda, Jahanabad, Patna, Kaimur, Arwal, Rohtas. | Sabour (Bhagalpur), BAU Sabour. |
| 4. | DAMUs in (Zone IIIA & IIIB) | Sheikhpura, Banka, Jamui, Gaya, Nawada, Aurangabad. | DAMUs (District Agromet Unit). |

## 5.7.1 Objectives of the project

The first objective of the project is to pass on weather data biweekly to NCMRWF, Delhi and DDGM (Agrimet) India Meteorological Department,

Pune from 1.8.2007. The second objective is to prepare agromet advisory bulletin on the basis of weather forecast received from Meteorological centre, Patna after modification of the forecast. Dissemination of the Agromet Advisory Bulletins (prepared biweekly) is the another objective through All India Radio, different TV channels, Newspapers, IMD website, Emails, whats app groups and also through personal contact to the selected farmers of zone III A & III B of Bihar. Regular collection of feedback from the selected farmers of these zones is also the duty of AMFUs. At meanwhile verification of the reliability of weather forecasts using actual observation recorded in the observatory is also one of the objectives of the project (Fig. 5.2). At First time, the preparation of Agromet Advisory Bulletin was started weekly from April 1997 and from 1998 it was started biweekly preparation of bulletin.

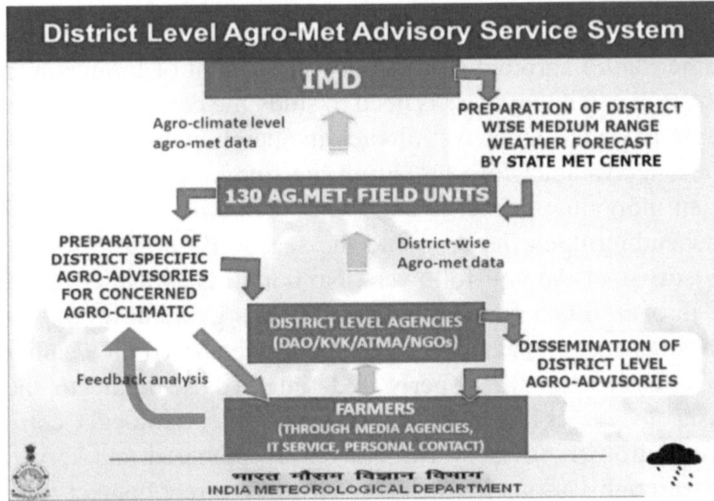


**Fig. 5.2:** District level Agromet Advisory Service System. (*Source*: IMD)

## 5.7.2 District Agromet Advisory Services by farmer's portal

Initially there were some problems in preparing District Agromet Advisory Bulletin through farmer's portal. It was needed fast internet connectivity for online preparation. So, it could not be prepared online through farmer's portal. In Hindi, it was also needed to improve the software so that some mistake in name of place or crop would be improved. But now a days it working properly and software has been improved.

## 5.7.3 Feedback from the farmers

In order to improve the quality of the agromet advisory services, regular direct interactions are being made with local farmers. The centre regularly participates in Kisan Chaupal (Gosthi), Kisan mela, Farmers gathering and interacts with farmers and collects feedback from the farmers. Surveys are conducted to obtain the feedback from the participating farmers. Data are collected through semi-structural questionnaire and personal contact is made with the farmers for getting the feedback from the participating farmers. As a case study, out of 50 farmers only 44 farmers replied to the questionnaire in Bihar under one district. It was seen that about 90% farmers were benefited through weather forecasts and Agromet Advisory Bulletins with respect to various agro management viz; timely preparation of land, use of suitable varieties, timely sowing of seeds, amount of water applied, timely irrigation scheduling, fertilizer application, seed treatment with suitable chemicals, plant protection measures and its application etc. The responses of the farmers were observed to be very positive and encouraging.

### 5.7.4 Impact of Agromet Advisory Bulletins

To strengthen weather based agromet advisory for the benefit of farmers and to ensure food security it is felt that there is need to study the economic impact of weather based agro advisory which was already in operation under National Centre for medium range weather forecast. From the study it is clear that the economic gain by an individual farmer is considerable and it varied from 10 to 35% or more depending upon the crop and the season if one follows the biweekly agro advisory over the non-followers. No doubt, the weather based agro advisory has an overall beneficial effect and farmers gain knowledge in addition to monetary benefit. The agromet advisory service based on weather forewarning is an effective tool if properly and timely disseminate to the farmers. On the basis of weather forecast received from Meteorological Centre (MC), Patna, Bihar, Agromet Advisory Bulletins were prepared and served among the farmers through various Medias as well as by direct contact with the local farmers of the region. A sample survey was made to get the feedback from the participating farmers. The views of the farmers are collected through semi-structural questionnaire regarding benefit of agromet advisory bulletins circulated, are being briefly presented below as a case study in Bhagalpur district of Bihar:-

During the month of April, 15, 2017 there was prediction of no rain and the same was communicated time to time to the farmers of this region with advise to continue their harvesting and threshing operations of *rabi* crops without any delay. Thus, the farmers could take advantage of clear weather and dry westerly winds in all these post-harvest operations of crops.

During the month of May, 15, 2017 there was forecast of no rain and the farmers were advised in this forecast to irrigate the summer crops and vegetables and for spray in chilli. Following our suggestions from the advisories, the farmers were thus benefited by good crops and vegetables.

During the month of June, 2017 there was forecast of rain between 12-14 June, accordingly farmers were advised to keep themselves in readiness for sowing of medium duration of rice varieties in the nursery. At the same time they were also advised to sow *kharif* fodder. The rain received the above said period and the farmers were benefited by timely sowing of *kharif* fodder and timely sowing of medium and long duration rice seedlings for raising seedlings.

During the month of July, 2017, there was prediction of heavy rain during 8-10 July and similar situation occurred during the above said period. The farmers were advised to reap the benefit of predicted rain in transplanting of long and mid duration rice and sowing of short duration rice for seedlings if not done earlier. Thus most of the farmers could reap the benefit of predicted rain in transplanting of rice and also utilized rain water in irrigation to prior transplanted rice crop.

During the month August, 2017 there was forecast of rain during 01-02 August. The same was communicated to the farmers of this locality and were advised to transplant short duration and photosensitive local tall varieties of rice seedlings, if not transplanted earlier. At the same time they were also advised to utilize predicted rain in irrigation to transplanted rice crop and to drain excess water from *kharif* maize, pulses and oilseed crops to avoid water logging. Rain occurred during the period and farmers were benefited by timely transplanting of rice crop, utilized rain water in irrigation to prior transplanted rice and saved their other Rabi crops from water logging.

During the month of September, 2017 there was forecast of no rain during 9-13 September 2015. Accordingly farmers were advised to irrigate the transplanted rice crop and to maintain the water in the field due to panicle initiation stage of the crop which is very crucial period for better production.

During the month of October, 2017 there was prediction of no rain during 17 to 21 October 2015 and farmers were advised to spray Quinalphos in rice crop due to attack of Gandhi bug. The same was followed by the farmers and took the benefit of timely spraying of the insecticide in rice crop.

During the month of November and December, 17 there were forecast of no rain and the same weather situations continued during that month. The farmers were communicated about the prediction of clear weather and advised to reap the benefit of clear weather in harvesting of matured rice crop and to prepare field for sowing of *rabi* crops like-maize, wheat, pulses and oilseeds. The same was followed by the farmers and took the benefit of timely sowing of above mentioned crops.

During the month of January, 2018 there were the prediction of no rain during 6-10 January. This was communicated to the farmers well in advance and was advised to irrigate the crop and spray insecticides and fungicides for control of insects, pest of their *rabi, maize*, potato and wheat crops. No rain was received during that said period and thus the adoptive farmers could save their crop from attack of pest.

During the month of February, 2018, there was prediction of no rain and the same was communicated time to time to the farmers of this region with advice to spray the insecticides in mustard against aphid and fungicides in pea against powdery mildew. Thus, the farmers could take advantage and sprayed the same to protect the above mentioned crops.

During the month of March, 2018, there was prediction of no rain and the same was communicated time to time to the farmers of this region with advice to spray the insecticide in chickpea against pod borer. Thus, the farmers could take advantage and sprayed the same to protect the mentioned crop.

### 5.7.5 Economic impact of Gramin Krishi Muasam Sewa

The benefits to the farmers by using Agromet Advisory Bulletins and weather forecast for taking farm level decisions have been discussed here.

#### 5.7.5.1 Case study I

The agro-advisory based on weather forecast, contains a summary of previous week's weather, deviation of weather from the normal value, weather forecast for next five days, suggestion of crop management practices based on weather forecast and warning to the farmers in advance regarding variation in rainfall or other weather parameters including pest and disease problems. Thus, farmers can choose suitable crop management options and strategies to overcome other problems. Weather forecast based agromet advisories helps in increasing the economic benefits to the farmers by suggesting them the suitable management practices according to the forecast weather condition. A case study was carried out on adaptation of agromet advisory services and its economic impact on wheat and carrot during *rabi* 2017-18 and on rice crop during *kharif* 2017. For assessing the impacts of agromet advisory services, users of agromet advisory services (AAS) and non-users of agromet advisory services (non AAS) were selected for wheat, carrot and rice crop. The study area lays around 100 km range from BAU, Sabour, Bhagalpur.

The study showed that the farmers who followed ASS were able to lower the input cost up to 6%, 9.6% and 7% in wheat, carrot and rice respectively and the net profit was increased by 0.9% in wheat, 3% in carrot and 4% in wheat as compared to the non AAS farmers, who did not follow ASS. AAS farmers were able to reduce the input cost up to Rs. 835/acre in wheat, Rs. 2616/acre in carrot and 1072/acre in rice. Increases in the net profit were Rs. 1040/acre in wheat, Rs. 4532/acre in carrot and Rs. 2214/acre in rice compared to the non AAS farmers. More net returns of AAS farmers over non-AAS farmers were gained due to low input cost by following weather based management practices such as timely land preparation and sowing, use of recommended seed rate and suitable varieties, timely weeding, harvesting and irrigation; and timely management of insect pests and diseases.

The study also showed that the application of current and forecasted weather based agromet advisories in different farm operations is a beneficial tool for enhancing the production of crops as well as income of the farmers. AAS farmers received weather based agro-advisories, including optimum and timely use of inputs for different farm operations. Production cost for the AAS farmers was reduced due to judicious and timely utilization of inputs. Increase in yield level and reduction in cost of cultivation led to increase net return.

### 5.7.5.2 Case study II

Gramin Krishi Mausam Sewa is a useful information tool to the farmers of this region in deciding their planning and budgeting for weather-agro management operations to achieve maximum benefit from predicted weather forecast served to them periodically by Agromet Advisory Services Unit comprising of eminent scientists of different disciplines of Bihar Agricultural College, Sabour.

The table 2 indicates that in case of *kharif* rice, maize and brinjal monetary value of saving was estimated as Rs. 2880, Rs. 3380 and Rs. 4,50,000 per hectare for respective crops. The estimated loss by non-adoptive farmers who did not follow the agro-advisories provided by Gramin Krishi Mausam Sewa (GKMS), served by this centre was estimated at 240, 260 and 03 MT per hectare in rice, maize and brinjal respectively under this region.

In case of *rabi* season maize, gram, lentil, wheat and cauliflower monetary value of saving was estimated Rs. 4060, Rs. 1600/-, Rs. 1750/-, Rs. 3000/- and 4,20,000/- per hectare respectively. The grain yield losses to maize, gram, lentil, wheat and cauliflower was 290 kg, 200 kg, 250 kg 300 kg and 03 MT per hectare respectively for non-adoptive farmers who did not follow the agro-advisories.

Similarly, in case of summer season okra, moong, mango and litchi the benefits were estimated as Rs. 2,40,000/-, 7200/-, 3,60,000/- and 9,10,000/- per hectare respectively to adoptive farmers of the locality.

So, it is concluded that the AAS adaptive farmers got benefit of Rs. 1600 to 4060 for cereal crops, while Rs. 2.4 lakh to 4.5 lakh and Rs. 3.6 to 9.1 lakh for vegetables and fruit crops respectively.

**Table 5.2:** Monetary gains accrued to farmers during the year 2017-18.

| Season | Crops grown by the farmers | Mean productivity realized in kg/ha | | Additional Production gained by adoptive farmers (kg/ha) | Price Rs./kg | Additional income Rs./ha |
|---|---|---|---|---|---|---|
| | | ASS adoptive farmers | ASS Non-adoptive farmers | | | |
| 1.Kharif | Rice | 3000 | 2760 | 240 | 12.00 | 2880/- |
| | Maize | 4000 | 3740 | 260 | 13.00 | 3380/- |
| | Vegetables (Brinjal) | 20 MT | 17 MT | 03 MT | 15 | 4,50,000/- |
| 2. Rabi | Wheat | 3500 | 3200 | 300 | 10.00 | 3000/- |
| | Maize | 5800 | 5510 | 290 | 14.00 | 4060/- |
| | Gram | 1200 | 1000 | 200 | 80.0 | 1600/- |
| | Lentil | 1150 | 900 | 250 | 70.0 | 1750/- |
| | Vegetable (Cauliflower) | 21 MT | 18 MT | 3 MT | 14 | 4,20,000/- |
| 3. Summer | Moong | 0880 | 0760 | 120 | 60.00 | 7200/- |
| | Vegetable (Okra) | 15 MT | 13.5 MT | 1.5 MT | 16.0 | 2,40,000/- |
| | Fruit1. Mango | 8.2 MT | 7.0 MT | 1.2 MT | 30 | 3,60,000/- |
| | 1. Litchi | 7.6 MT | 6.3 MT | 1.3 MT | 70 | 9,10,000/- |

## 5.8 Verification and Analysis of Weather Forecast

A five day medium range weather forecast was received from Meteorological Centre (MC), Patna on every Tuesday and Friday of the week. The weather forecast data of whole year was grouped into four seasons i.e. pre-monsoon, monsoon, post-monsoon and winter season for analysis and verification of the forecast data. Both qualitative and quantitative verification analysis were carried out using skill score and critical values for error structure. The correlation co-efficient and root mean squire error were also worked out for all the four seasons.

### 5.8.1 Qualitative verification analysis

#### 5.8.1.1 Rainfall forecast verification

Skill score test, which is based on 2 X 2 contingencies table, is used for verification of qualitative analysis of rainfall forecast as suggested by

NCMRWF. The results for all four seasons for the year 2017-18 have been presented in table 5.3.

It is found from the table 3 that the value of ratio score was the highest during post-monsoon season with ration score value of95%, followed by pre-monsoon (94%) and winter season (93%). The ratio score value during monsoon season was about 68% which clearly shows that there was better accuracy of rainfall forecast during monsoon season.

The threat score, which considers only YY cases, is also found maximum during pre-monsoon season with the value of 82%. While, during monsoon, post monsoon and winter seasons threat score values were 53, 52 and 0 % respectively.

**Table 5.3:** Trends of rainfall prediction during different seasons at AMFU, Sabour under Bihar Agricultural College, Sabour (Year 2017-18).

| S. No. | Type of skill score | SEASON | | | |
|---|---|---|---|---|---|
| | | Pre-Monsoon | Monsoon | Post-Monsoon | Winter |
| 1. | Ratio score | 0.943 | 0.679 | 0.952 | 0.932 |
| 2. | Bias score | 0.921 | 0.604 | 0.523 | 0.677 |
| 3. | Probability of detection | 0.867 | 0.572 | 0.543 | 0.000 |
| 4. | False alarm ratio | 0.015 | 0.036 | 0.000 | 0.032 |
| 5. | Threat score | 0.823 | 0.536 | 0.521 | 0.000 |
| 6. | Haidke skill score | 0.778 | 0.468 | 0.391 | -0.684 |
| 7. | Hansen & Kuipper score | 0.843 | 0.402 | 0.506 | -0.036 |

## 5.8.1.2 Analysis verification of other weather parameters

Qualitative analysis verifications of cloud cover, wind speed and direction, maximum and minimum temperature were also carried out using standard statistical procedure for all the four seasons and have been presented in table 5.4.

**Table 5.4:** Season wise correlation co-efficient (CC) and root mean square error (RMSE) values of different weather parameters.

| S. No. | Weather parameter | Season (2017-18) | | | | | | | |
|---|---|---|---|---|---|---|---|---|---|
| | | Pre-Monsoon | | Monsoon | | Post-Monsoon | | Winter | |
| | | CC | RMSE | CC | RMSE | CC | RMSE | CC | RMSE |
| 1. | Cloud cover | 0.549 | 2.332 | 0.442 | 2. 778 | 0.5467 | 2.794 | 0.561 | 2.066 |
| 2. | Rainfall | 0.802 | 4.966 | 0.309 | 15.315 | 0.9400 | 9.021 | 0.802 | 0.880 |
| 3. | Wind speed | 0.720 | 2.863 | 0.771 | 3.749 | 1.1354 | 3.548 | 0.426 | 3.739 |
| 4. | Wind direction | 0.545 | 98.678 | 0.359 | 82.287 | 0.3592 | 90.125 | 0.568 | 95.522 |
| 5. | Max. temp. | 0.877 | 1.651 | 0.771 | 2.112 | 0.9665 | 1.055 | 0.837 | 2.825 |
| 6. | Min. temp. | 0.870 | 2.378 | 0.230 | 1.825 | 0.9453 | 2.197 | 0.781 | 2.672 |

#### 5.8.1.3 Correlation-coefficient and Root Mean Square Error (RMSE)

The perusal of correlation coefficient and root mean square errors data which were worked out using standard statistical procedure between weather forecast and actual weather prevailed during the same period indicated that the forecasts made by this AMFU were more or less close to correctness excluding wind direction. All observed weather parameters via; cloud cover, rainfall, wind speed, max. & min. temp, except wind direction were found in the line of forecast made in all the four seasons respectively. The RMSE values of wind direction were found too high in all the four seasons to accept any homogeneity in the predicted and observed values. The RMSE value of rainfall during monsoon season was also higher which clearly indicated that forecasts of rain were more or less correct but amount of rain predicted never tallied with observed value of rain occurred.

### 5.8.2 Quantitative verifications analysis

The quantitative verification analysis worked out between weather forecast made and actual weather prevailed during the same period and has been presented in table 5.5 (a, b, c, and d). Total numbers of forecasts received during the year 2017-18 were 360 out of which 91 were during pre-monsoon season, 121 were during monsoon season, 92 were during post-monsoon season and 56 were during winter season. These forecasts have been verified by correct, usable and unusable methodology. In this procedure some limits of predicted values of different parameters as suggested by NCMRWF, were used for quantitative verification analysis. Forecast values falling within these limits are recorded as correct and usable and beyond these limit, the forecasts are rated as unusable for meteorological application.

#### 5.8.2.1 Cloud cover

From the above tables it is evinced that out of 91, 121, 92 and 56 cloud cover forecasts received during pre-monsoon, monsoon, post-monsoon and winter season respectively; 85, 86, 75 and 82 % were correct in the respective season. The accuracy of cloud cover forecast ranged between 75 to 86 % during the year 2017-18.

#### 5.8.2.2 Rainfall

Total number of rainfall forecast received during pre-monsoon, monsoon, post-monsoon and winter season was 91, 121, 92 and 56 respectively. While 95, 74, 97 and 93 % respectively were correct. During the rainy season the accuracy of rainfall forecast was only 69 %, while during other seasons the accuracy of rainfall forecast was much higher because on most of the days there were neither prediction of rainfall nor it occurred during the said period.

### 5.8.2.3 Wind speed

Total number of wind speed forecast received during the four meteorological seasons i.e. pre-monsoon, monsoon, post-monsoon and winter was 91, 121, 92 and 56 respectively. Out of these wind speed forecasts 64, 49, 51 and 79 % were found correct in the respective season. The accuracy percent of wind speed forecast was highest (79%) during winter season and lowest (49%) during monsoon season.

### 5.8.2.4 Wind direction

In case of wind direction, out of 91, 121, 92 and 56 forecasts received during pre-monsoon, monsoon, post-monsoon and winter season respectively; 51% during pre-monsoon, 49% during monsoon, 71% during both post-monsoon and winter season were found correct, which indicates that on most of the days forecast of wind direction was beyond and specified limit of ± 30$^0$.

### 5.8.2.5 Maximum temperatures

Total number of maximum temperature forecast received during pre-monsoon, monsoon, post-monsoon and winter season was 91, 121, 92 and 56 respectively. While 82, 93, 87 and 43 received forecasts were correct during the respective season. This clearly indicated that the accuracy percentage of maximum temperature forecast was 90, 77, 95 and 76 in the respective season.

### 5.8.2.6 Minimum temperature

Similarly in case of minimum temperature, out of 91, 121, 92 and 56 forecasts received during the respective season; 81, 96, 74 and 47 received forecasts were correct indicating 88, 79, 81 and 84 % accuracy in the respective season.

**Table 5.5 (a):** Weather parameters during pre-monsoon season, 2017.

| S. No. | Weather parameter | Pre-monsoon Season, 2017 | | | |
|---|---|---|---|---|---|
| | | Correct | Usable | Unusable | Total |
| 1. | Cloud cover | 66 (72 ) | 12 (13 ) | 13 (14 ) | 91 (100 ) |
| 2. | Rainfall | 84 (92 ) | 03 (3 ) | 04 (4 ) | 91 (100 ) |
| 3. | Wind speed | 45 (49 ) | 14 (15 ) | 32 (35 ) | 91 (100 ) |
| 4. | Wind direction | 34 (37 ) | 16 (17 ) | 41 (45 ) | 91 (100 ) |
| 5. | Maximum Temperature | 72 (79 ) | 10 (11 ) | 09 (10 ) | 91 (100 ) |
| 6. | Minimum Temperature | 67 (73 ) | 14 (15 ) | 10 (11 ) | 91 (100 ) |

**Table 5.5 (b):** Weather parameters during monsoon season, 2017.

| S. No. | Weather parameter | Monsoon Season, 2017 | | | |
|---|---|---|---|---|---|
| | | Correct | Usable | Unusable | Total |
| 1. | Cloud cover | 88 (72) | 17 (14) | 16 (13) | 121 (100) |
| 2. | Rainfall | 65 (53) | 26 (21) | 30 (25) | 121 (100) |
| 3. | Wind speed | 32 (26) | 23 (19) | 66 (55) | 121 (100) |
| 4. | Wind direction | 48 (39) | 12 (10) | 61 (51) | 121 (100) |
| 5. | Maximum Temperature | 57 (47) | 36 (30) | 28 (23) | 121 (100) |
| 6. | Minimum Temperature | 59 (47) | 37 (32) | 26 (21) | 121 (100) |

**Table 5.5 (c):** Weather parameters during post monsoon season, 2017.

| S. No. | Weather parameter | Post-monsoon Season, 2017 | | | |
|---|---|---|---|---|---|
| | | Correct | Usable | Unusable | Total |
| 1. | Cloud cover | 61 (66) | 08 (9) | 23 (25) | 92 (100) |
| 2. | Rainfall | 88 (96) | 01 (1) | 03 (3) | 92 (100) |
| 3. | Wind speed | 35 (38) | 12 (13) | 45 (49) | 92 (100) |
| 4. | Wind direction | 46 (50) | 19 (21) | 27 (29) | 92 (100) |
| 5. | Maximum Temperature | 60 (65) | 27 30) | 05 (5) | 92 (100) |
| 6. | Minimum Temperature | 53 (58) | 21 (23) | 18 (19) | 92 (100) |

**Table 5.5 (d):** Weather parameters during winter season (2017-18).

| S. No. | Weather parameter | Winter Season (2017-18) | | | |
|---|---|---|---|---|---|
| | | Correct | Usable | Unusable | Total |
| 1. | Cloud cover | 40 (71) | 6 (11) | 10 (18) | 56(100) |
| 2. | Rainfall | 46 (82) | 6(11) | 4 (7) | 56(100) |
| 3. | Wind speed | 29 (52) | 15 (27) | 12 (21) | 56(100) |
| 4. | Wind direction | 23 (41) | 17 (30) | 16 (29) | 56(100) |
| 5. | Maximum Temperature | 32 (57) | 11 (19) | 13 (24) | 56(100) |
| 6. | Minimum Temperature | 33 (59) | 14 (25) | 9 (16) | 56(100) |

Fig. in ( ) indicates in per cent.

## 5.9 Agromet Advisory Bulletin Based on NDVI Map

The map (Fig. 5.3) is regularly prepared by India Meteorological Department and is sent at fortnightly to different AMFUs for preparation of Agromet Advisory Bulletin. Agriculture vigour is observed for the period and for the area concerned. The analysis is done based on the vigour and weather. Then Agromet advisory bulletin is prepared for the next five days based on the study.

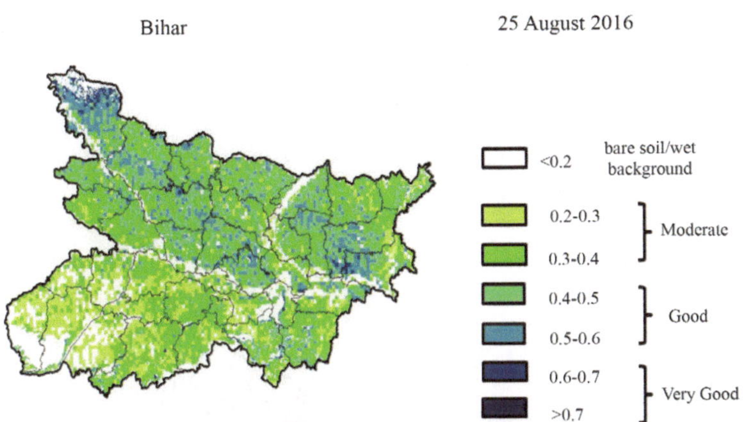


**Fig. 5.3:** NDVI (Normalized Difference Vegetation Index) map.(*Source*: Shibendu S. Ray, 2016)

## 5.10. Agromet Adviosry Bulletin Based on the Standardized Precipitation Index (SPI) Map

The map is received from IMD Pune (Fig. 5.4). This is prepared based on the rainfall and its deviation from the normal of the area and the period. The map with full details of the map is sent to the AMFUs by e mail.

### 5.10.1 Details about the map of SPI (Standardized Precipitation Index)

Most parts of Maharashtra, Madhya Pradesh, Uttar Pradesh, some parts of Andhra Pradesh, Karnataka, Rajasthan; Rayagada and Udhamsinghnagar district in Uttarakhand; Bilaspur district of Himachal Pradesh; Patna, Rohtas Munger districts of Bihar; Nawarangpur and Koraput districts of Odisha; Durg, Koriya, Surguja, Janjgir, Raipur districts in Chattisgarh; Gurgaon and Bhatinda districts of Punjab; Faridabad, Rewari, Chandigarh districts of Haryana; Pulwama and Reasi districts in Jammu and Kashmir; Dangs, Junagarh, Porbhandar, Navsari, Narmada, Diu districts in Gujarat, Saurashtra and Kutch experienced extremely wet or severely wet conditions (Fig. 5.3).

Moderately or severely dry conditions experienced over South Tripura district of Tripura; Bongaigaon, Karimganj, Lakimpur, Sonitpur, Sibsagar, Tinsukia, Kamrup Metro districts of Assam; East Siang and Papumpara districts of Arunachal Pradesh.

Extremely dry conditions experienced over Ladakh and Poonch districts of Jammu & Kashmir; Dibrugarh, Golaghat Jorhat, Morigaon districts of Assam; Changlang, Lohit, East Kameng and West Kameng districts of Arunachal Pradesh. Rest of the country experienced mild wet or mild dry conditions.

MINISTRY OF EARTH SCIENCES
INDIA METEOROLOGICAL DEPARTMENT
HYDROMET SECTION, PUNE

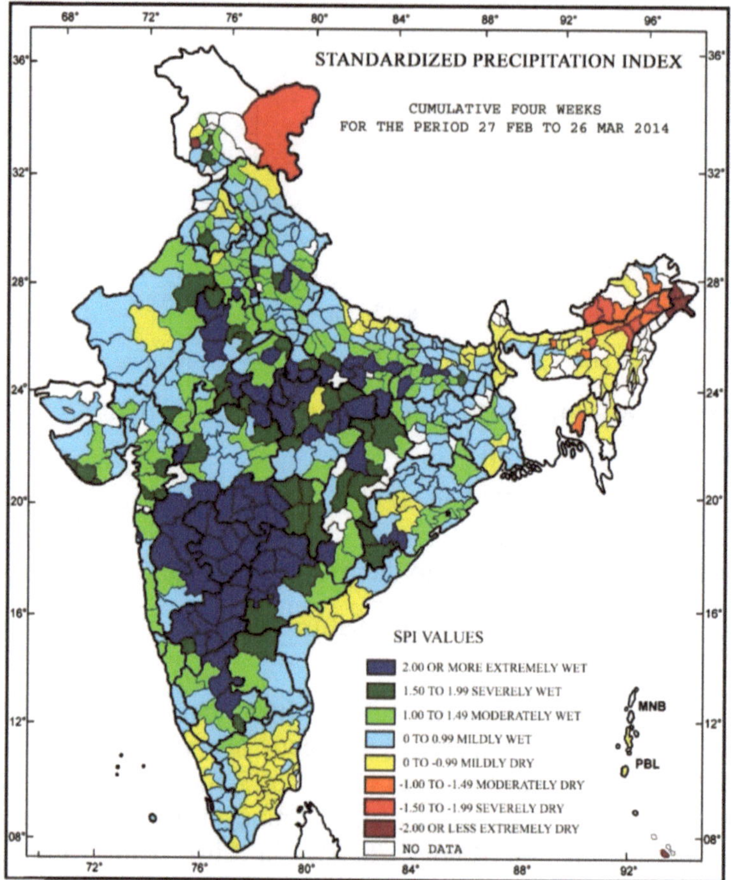


**Fig. 5.4:** Standardized Precipitation Index from 27 February to 26 March 2014. (Source: IMD)

The Standardized Precipitation Index (SPI) is prepared by Hydromet section of India Meteorological Department (IMD) and sent fortnightly to every AMFU centre of the Country. Considering the moisture condition of the soil, advisory regarding irrigation is provided to the farmers based on the weather forecast for upcoming five days.

## References

Damrath U., Doms G., Friihwald D., Heise E., Richter B. and Steppeler J. 2000. Operational quantitative precipitation forecasting at the German Weather Service. J of Hydrology. 239, 260-285.

Rathore L.S., Gupta Akhilesh and Singh K.K. 2001. Medium range weather forecasting and agricultural production. Journal of Agric. Physics. 1(1), 43.

Tamiotti L., Olhoff A., The R., Sommons B., Kulacoglu V. and Abaza H. 2009. Trade and climate change: a report by the United Nations Environment Programme and the World Trade Organization. WTO, Geneva, Switzerland.

UNEP, 2011. Decoupling natural resource use and environmental impacts from economic growth. A Report of the Working Group on Decoupling to the International Resource Panel. Fischer-Kowalski, M., Swilling, M., von Weizsäcker, E.U., Ren, Y., Moriguchi, Y., Crane, W., Krausmann, F., Eisenmenger, N., Giljum, S., Hennicke, P., Romero Lankao, P., Siriban Manalang, A. United Nations Environment Programme.

World Bank, 2010a. World development report 2010: development and climate change. The World Bank, Washington, DC, USA.

# 6

# Agromet Advisory to Mitigate Impact of Climate Change

## 6.1 Introduction

One of the key challenges in the 21st century is that Agromet Advisory Services (AAS) have to respond to an unpredictable and changing climate. In India Gramin Krishi Mausam Sewa (GKMS) is the umbrella project for Agromet Advisory Services (AAS) and it intends to link AAS to different sources of knowledge and innovations to respond to the climate change at root level farmers. The main aim of GKMS is to create effective, efficient and synergistic linkages to improve the delivery of these AAS to the farmers. GKMS works under the frame work of India Meteorological Department (IMD) which seeks to enhance the livelihood of every Indian farmer. It directly focuses on the needs of the farmers, contributing to sustainable growth in and transformation of Indian agriculture by providing weather based effective advisory services. It introduces the opportunities and challenges for AAS in response to changing climate and begins to outline the possible roles of adaptive AAS.

## 6.2 Understanding the Linkages Between Climate Change Agriculture and Advisory Services

### 6.2.1 Key issues in climate change and agriculture

#### 6.2.1.1 *Climate change*

It is widely accepted that human's day-to-day activities are increasing the level of different greenhouse gases (GHGs) in the atmosphere, resulting global warming, which is leading to changes in the climate. The Intergovernmental Panel on Climate Change report (IPCC, 2007) defined that: '*Warming of the climate system is unequivocal, as is now evident from observations of increases in global average air and ocean temperatures, widespread melting of snow and ice and rising global average sea level*' and '*Most of the observed increase in global average temperatures since the mid-20th century is very likely due to the observed increase in anthropogenic GHG concentrations*'.

Climate can be defined as the 'long term average weather' (IPCC, 2007). IPCC defined 'climate change' as 'any change in climate over time, whether due to natural variability or as a result of human activity'. The United Nations Framework Convention on Climate Change (UNFCCC) described climate change as 'a change of climate which is attributed directly or indirectly to human activity that alters the composition of the global atmosphere and which is in addition to natural climate variability observed over comparable time periods'. Future warming will highly dependent on natural factors and human impact on the level of GHG emission into the atmosphere. These GHGs are mainly carbon dioxide ($CO_2$), methane ($CH_4$) and nitrous oxide ($N_2O$). Most of the scientists agree that global warming will continue through the 21st century and we are already committed to a certain level due to historical emissions. It is impossible to predict the certain changes that will occur in the climate. The complexity and difficulty in modelling and insufficient climate data are the limitations to make assumptions about future decisions. Various scenarios or storylines have been developed by the IPCC, representing different demographic, social, economic, environmental and technological developments. Even an idealised situation of GHG concentration being held to the levels of 2000, it is estimated that still global temperature will rise on an average of 0.6°C by the end of the 21st century compared to the end of the 20th century (IPCC, 2007). A rise of 1.8°C is the best estimate for the low emission scenario and a staggering rise of 4°C is the best estimate for the high emission scenario. These changes have catastrophic implications for humanity (IPCC, 2007). Changes in air temperature will also affect precipitation and wind patterns, frequency and intensity of extreme weather events and sea ice. However, the changes in climate will vary significantly with location.

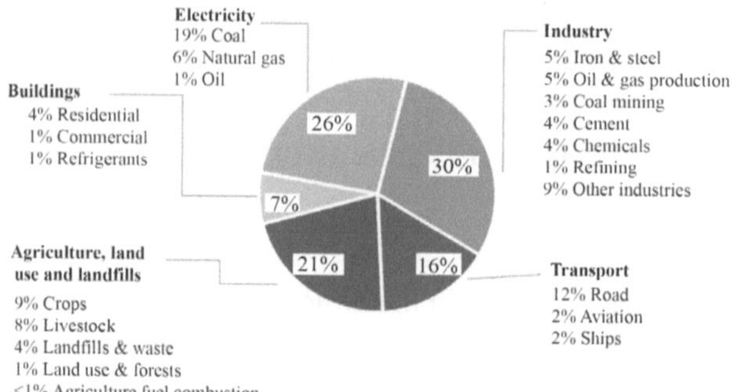


**Fig. 6.1:** Total anthropogenic greenhouse gas (GHG) emissions from economic sectors in 2019. (*Source*: Rhodium Group)

### 6.2.1.2 Climate change and agriculture

Agriculture is vulnerable to climate change, but agricultural sectors are also directly and indirectly a major contributor to GHG emission as well as to global warming. Further warming will create flourishing negative impacts in all regions. In several ways climate change may affect smallholder agriculture (Morton, 2007). Four direct impacts are getting more concerned– impacts on: (i) biological processes creating harmful effects on crops and animals; (ii) environmental and physical processes affecting production at watershed, landscape or community level; (iii) human health; and (iv) non-agricultural livelihoods. There also have some indirect impacts including: (i) off site impacts and (ii) impacts on adaptation and mitigation strategies. Global food production potential is likely to increase with the rise in global average temperature up to 3°C, but above this it is likely to decrease in the developing countries (Parry *et al.,* 2007). Inter-Governmental Panel on Climate change (IPCC) in its fifth assessment report described that increasing number of warm days and decreasing number of cold days have been observed with the warming trend continuing into the new millennium. Increasing trends of annual mean temperature in East and South Asia have been observed during the 20$^{th}$ century. Mean change in mean annual temperature under RCP8.5 model exceed 2°C during the late 20$^{th}$ century, 3°C in the mid of 21$^{st}$ century over South and Southeast Asia and 6°C in the late 21$^{st}$ century over high latitudes. Precipitation trends including extreme rainfall days show strong variability with both increasing and decreasing trends in different parts and seasons of Asia. In South Asia, mean rainfalls during different seasons show inter decadal variability. Noticeably a decreasing rainfall trend with more frequent monsoon deficit under regional level is homogeneous. Increase in the number of monsoon break days and decline in the number of monsoon depressions are continuing with the overall decrease in seasonal mean rainfall over India (IPCC, 2014). General Circulation Models (GCMs) and Special Report on Emission Scenarios (SRES), show that higher temperature causes lower yield of rice as a result of shorter growing periods. Several regions are already near the heat stress limits for rice cultivation. However, carbon dioxide fertilization may at least offset some yield losses in rice and other crops. About 50% decrease have been seen in the most favourable and high yielding wheat growing areas in the Indo-Gangetic Plains of South Asia as a result of heat stress (IPCC, 2014).

**Table 6.1:** Types of impacts of climate change on smallholder and subsistence agriculture.

| **Direct climate change impacts on smallholder livelihoods.** | |
|---|---|
| Biological processes which affect crops and animals at the level of individual organisms or fields | Direct impacts of changes in carbon dioxide, temperature and precipitation on yields of cash crops, productivity and health of livestock.<br>Can include impacts of variability in temperature and precipitation, eg, hot or dry spells at key stages in crop development. Also includes changed patterns of pests and diseases. |

| | |
|---|---|
| Impacts of climate change on human health | The above mentioned impacts of climate change on agriculture will be combined with impacts on human health and the ability to provide labour for agriculture. |
| Impacts of climate change on non-farm livelihoods | Impacts on secondary non-agricultural livelihood strategies like tourism for many rural people in developing countries. |
| **Secondary or indirect impacts of climate change** | |
| Distant or off-site impacts of climate change on a particular smallholder system | Impacts of climate change on other distant areas may create changes which may affect a smallholder system. For example, decreased supply of grain in one location might affect specialist cash-crop producers in another area as the latter are net grain buyers. |
| Impacts of adaptation and mitigation policies for climate change, programmes and funds | The secondary impacts occur as governments, civil society, the private sector, etc., gear up to respond to climate change and institute new policies, programmes and funds – all of which may impact upon smallholders (positively or negatively). An example would be leasing of agricultural lands to agribusiness for biofuel production. |

*Source*: Adapted from Morton (2007).

The number of people at risk of hunger due to climate change depends on overall socio-economic development on the country. Small and marginal farmers will be suffered more by localised impacts of climate change (Parry *et al.,* 2007). Modelling the possible impacts of climate change is very complex and there are also some uncertainties. Different uncertainties make the process to estimate the impacts of climate change on agriculture more challenging (Betts *et al.*, 2009). These uncertainties include: $CO_2$ fertilization (i.e., physiological response of the crop plants to atmospheric $CO_2$ concentration and its impact on yields); crop sensitivity (i.e., the sensitivity of different crops to local level climate change); uncertainties in climate models (eg, the different emission scenarios); and regional rainfall patterns (there is less agreement among climate models in projection of regional rainfall pattern than that of temperature).

**Impact of climate change on agricultural production:** In most of the regions, crop production is likely be severely affected by climate change and climate variability. This is also adversely affecting food security and exacerbating malnutrition. Most of the livelihoods are reliant on agricultural yields and natural resources. Agriculture is one of the major contributors to Indian economy. Losses in agricultural production are possibly severe in several areas, accompanied by changes in length of growing periods affecting, arid, semi-arid and rainfed based cropping systems under certain climate projections. Yields in rainfed agriculture have reduced up to 50% by 2020 in some areas. Local people suffer from additional losses when climate change interacts with other weather hazards and stresses.

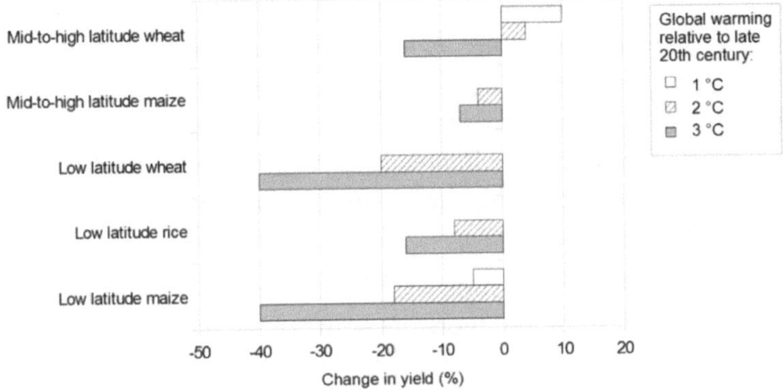


**Fig. 6.2:** Projected changes in crop yields at different latitudes with global warming. This graph is based on several studies. Source: The US National Research Council (*Source*: US NRC, 2011).

**Increase in aridity:** Due to drought and land degradation, agricultural yield is declining especially in marginal areas. Under various scenarios it has been noted that length of growing period is also changing.

**Pressure on water resources:** Currently water stress is likely to be increased in many areas by climate change and variability. Climate change is also reducing primary production of fish.

**Impact on ecosystems:** Indian ecosystem is likely to experience major shifts and changes in species and possible extinctions.

**Sea-level rises:** It is projected that rise in sea surface level will affect low lying coastal areas with large population towards the end of the 21st century. It is expected that the cost of adaptation will exceed 5-10% of GDP (Boko *et al.*, 2007).

These scenarios make different assumptions regarding GHG emission level, future economic growth and population. These results show increase in temperature associated with effects on yields of cereal crops and the number of people at risk of hunger. There are different ideas regarding the potential role of smallholder agriculture in pathways of future development. Every option and potential pathway for agricultural development must be re-assessed in the light of the challenges due to climate change.

### 6.2.2 Responding to climate change

Responses to climate change are mainly divided into two categories: adaptation (addressing effects) and mitigation (addressing causes).

#### 6.2.2.1 Adaptation to climate change in agriculture

Adaptation can be described as adjustment in natural or human system in response to actual or expected climatic condition and its effects, which moderates harmful effects and exploits beneficial opportunities. Different types of adaptation can be described as anticipatory, autonomous and planned adaptation (IPCC, 2007). Adaption to the weather or climate is a human characteristic, but every day climate change is presenting new challenges. Adaptation to climate change is growing up as an international development agenda (Nelson *et al.,* 2008).

Changes at farm level include modification of different farm practices aiming maintenance of the existing system, but there are some challenges like broader inequalities in land distribution, which may be more significant and systemic in nature. Changes in governance may also be needed to create an enabling environment for adaptation.

#### 6.2.2.2 The contribution of agriculture to climate change and mitigation strategies

Climate change is not only affecting agriculture, but agriculture is also one of the major contributors to the climate change. The agricultural sector is a significant source of GHGs, which contribute to the global warming. Though, agriculture has the potential to mitigation the climate change through: (a) reducing GHG emission, (b) enhancing removal of carbon (storing or sequestering or capturing), and (c) avoiding fossil derived emissions through production of bio fuel. Expansion of new mitigation strategies for livestock and fertiliser application is essential to prevent an increase in GHG emission from agricultural sector after 2030. The most promising options for mitigating GHG emission in agriculture include:

- Improved crop management practices (eg, improved agronomic practices, tillage and residue management, organic nutrient and pesticide use)
- Restoration of organic soils which are best for crop production, and also restoration of degraded lands.
- Improvement of water and rice management practices.

### 6.2.3 The contribution of agriculture to climate change

The agriculture sector alone contributed about 10-12% of total global anthropogenic GHGs emission (5.1-6.1 Gt $CO_2$ per year) in 2005, including about 50% of global anthropogenic methane emission (methane contributed 3.3 Gt $CO_2$ in total), about 60% of nitrous oxide (nitrous oxide contributed

2.8 Gt $CO_2$ in total). Emission rate of GHGs may increase in the future due to increasing population growth and changing diet. Higher population will generate greater demand for food resulting higher emissions of methane and nitrous oxide due to more livestock production and higher use of nitrogen fertilizers. The global technical mitigation potential from agricultural sector is estimated to be about 5500 - 6000 Mt $CO_2$ by 2030. A key determinant of how much of this potential is converted into action is the price of carbon (Smith *et al.*, 2007). Many mitigation practices based on existing technologies should be implemented immediately, but technological development will be a major factor influencing the efficacy of additional mitigation measures in the future.

Soil carbon sequestration offers most of the mitigation potential and shares almost 89% contribution to the technical potential. Methane and nitrous oxide emission from soil contribute about 9% and 2% to the total mitigation potential respectively (Smith *et al.,* 2007). As like adaptation strategies, there is no universally applicable list of mitigation practices. All mitigation practices need to be evaluated for appropriate use in individual agricultural system on the basis of climate condition, soil related factors, historical patterns of land use and management. The price of carbon is the key determinant of mitigation strategies. At low price farmers may adjust different production practices such as tillage operation, fertiliser application, livestock diet and manure management. Higher prices are needed to provide sufficient incentives for major land use changes. Agricultural mitigation practices often have synergy with sustainable development policies. Mitigation and adaptation strategies in agriculture can overlap, but macro-economic, agricultural and environmental policies must have greater impact on agricultural mitigation options than explicit climate policies. Despite of having significant technical potential for mitigation in agriculture, there is relatively little progress in the implementation of mitigation measures. Barriers of implementation are not be overcome without clear incentives and the solving of other issues related to agricultural innovation system such as strengthening the capacity of farmers, AAS and other actors. Most studies on AAS, agriculture and climate change have given little consideration to the different pathways like narratives, models and visions that are possible in agricultural development and the of agriculture. These economic models and ideas about the roles of smallholders in agriculture have to be analysed in the light of the challenges coming out from the climate change as well as other pressures and drivers. Major challenges for decision makers are to understand the context and potential adaptive strategies and pathways that are useful for farmers and other stakeholders in diverse rural level agriculture (Leach *et al.*, 2010) and to consider the implications of different paths of development. AAS have a key role to play in this context.

In the presence of climate change and other demands upon agriculture, a set of potential synergies and trade-offs in agricultural production, adaptation and mitigation strategies can be distinguished.

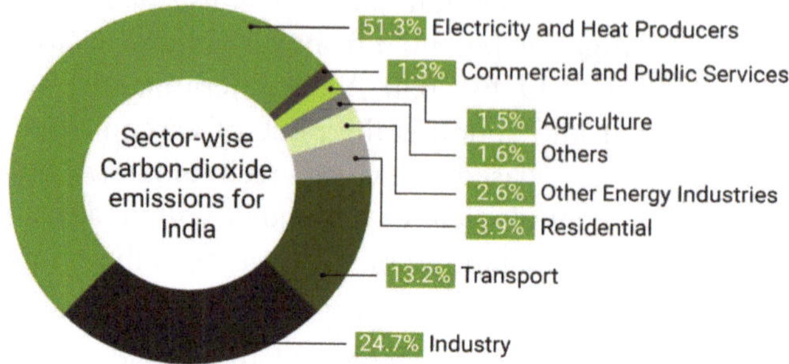


**Fig. 6.3:** Sector-wise $CO_2$ emission share in India.
(*Source*: $CO_2$ Emissions from Fuel Combustion, IEA 2021)

### 6.2.2 Adaptive agricultural advisory services

What should the new or expanded roles of AAS be in response to climate change and increasing demands on agriculture? With respect to the climate change, roles of AAS may need to include:

- Improvement of farmers' and other actors' access and use of indigenous weather and climate information which include climate science, as well as other forms of weather and climate knowledge (eg, local observations and adaptations).
- Analysis of the impacts of climate change and other drivers on farmer vulnerability and resilience to plan future responses.
- Strengthening the adaptive capacity and resilience of farmers and other AIS actors.
- Offering low-carbon development and climate mitigation services to other actors.

To fulfil these new or expanded roles, AAS will have to be developed in three ways (Christoplos, 2010b):

- **Capacity development to manage uncertainty:** Extension workers act as 'expert' providers of knowledge from researchers to the farmers and move for providing information, facilitation and advice related to possible trends and probabilities. Management of risk and uncertainty requires an improved understanding of climate change as well as technical, market and social uncertainties. AAS practitioners need greater capabilities in brokering information, innovating, facilitating and advising on probabilities and trends. More efforts should be needed to explore the possible future in rural areas over longer timescales in planning.

- **Respond to climate change and unpredictability** by helping farmers living with risk and less opportunities to adapt and transform livelihoods. This involves being more flexible and adaptive, responding in an integrated manner helping clients live with risk, and enabling farmers to identify and take up new opportunities; and to adapt and transform livelihoods.
- **Embrace pluralist extension systems** in the sense of the diversity of motivations and orientations of different types of providers (Christoplos, 2010b). Roles will vary depending on focus on training farmers, production, improving yields and technology transfer to facilitate the farmers and moving beyond training to learning. Other roles require more attention and will support farmer organisations, marketing issues, linking to other service providers and supporting farmers in their advocacy activities. GKMS should become more adaptive, i.e., they need to be able to adapt to climate change. There are some major principles of adaptive management. On the basis of these principles, AAS might evolve and respond to climate change. Adaptive mechanisms are established by explicit learning from policy experiments and by the use of new scientific information and technical knowledge to inform future decisions, improve understanding, monitor the outcome of interventions and also to develop new practices. Mechanisms are developed to enable the following points:
- Evaluation of alternative scenarios with structural and non-structural measures.
- Understanding and challenging the assumptions.
- Circumstantial consideration of uncertainties.
- Adoption of long term measures for strengthening and planning of capacity.
- Classification with ecological processes at appropriate scales.
- Frameworks for co-operation at administrative levels between sectors and line departments.
- Participation of broad stakeholders (including different research centres and NGOs) in decision making and problem solving.
- Enactment is adaptable to support local level action and respond to new information.

Adaptive AAS uses an 'adaptive management approach' that includes a shift in roles and outlook. AAS individuals, organisations and systems may be considered adaptive in terms of the extent to which they are:

- Enabling the farmers to rise up their assets to respond to climate change.
- Supporting equitable access to resources, especially by the most vulnerable.
- Providing support to farmers' self-organisation or agency in the light of challenges due to climate change.
- Use of technological innovation at farm and institutional innovation at policy level for adaptation and mitigation.
- Strengthening AAS by learning climate knowledge from farmers.
- Moving towards adaptive management:
    - Basic decisions on explicit learning from policy experiments and the use of new technical knowledge, scientific information and farmer knowledge to improve understanding, monitor the outcome of interventions, inform future decisions and to develop new practices
    - Longer timescale in planning and capacity strengthening.
    - Circumstantial consideration of uncertainties.
    - Evaluation of alternative scenarios with structural and non-structural measures.
    - Understanding and challenging the assumptions.
    - Classification with ecological processes at appropriate scales.
    - Having frameworks for cooperation at administrative levels between sectors and departments (for more integrated approaches).

## 6.3 Trends in AAS in the Light of Climate Change and Other Demands on Agriculture

### 6.3.1 Evolving roles and trends in AAS

Interpretations of AAS and extension services are diverse and have evolved over time. Now the role of extension has changed from a service that spreads research based knowledge to the rural sectors to improve farmers' livelihoods based on technology transfer, rural development, management skills and non-formal education, to a role of facilitation that provide learning and support to farmer groups on marketing and linking to a broader range of service providers and agencies (Davis, 2009).

Now focus is giving on AAS because of the increasing global concerns regarding climate change, agricultural productivity, pressure on land use, food price and security, oil prices and supply. Other discussions of extension (Swanson, 2008; Christopolos, 2010b) note the broader range of actors that are already or could be involved in AAS and the more multi-directional flow of information that could or should influence research programmes and agendas. The study includes the farmer field school approach and the agricultural technology management agency model of India. The attributes of farm households are also relevant in terms of their capacity, adoption of innovations, decision making and changes to practices. This contributes to the ultimate impacts of AAS in its context in terms of positive changes in yield, productivity, income, employment, empowerment, gender-specific impact, innovation, environmental effects, distributional effects and strengthening of value chains.

### 6.3.2 Population

Many researchers suggest that sustainable long term food production needs to double globally in order to meet the basic needs of increasing population. Demographic changes include not only urbanisation, but also increasing migration and seasonal mobility. Among the many factors, climate change and increasing population are highlighting the importance of different ecosystems and land resources. Climate change will make it harder to produce enough food for the world's growing population and will alter the availability and quality of water resources. To avoid expansion into other ecosystems, agricultural productivity needs to increase, while minimising the environmental damage and with net reduction in GHG emission from +farm operationsand postharvest activities (World Bank, 2010a). The concept of increasing output by using less resources and reducing the environmental pollution has been described by UNEP (2011) as 'decoupling'. Climate change is likely to alter the patterns of international trade by altering countries' comparative advantages in agriculture (Tamiotti *et al.*, 2009).

ASS includes a number of climate resilient practices such as small water infrastructure, watershed management, land use planning, intercropping, agroforestry, low tillage and soil erosion control that have advanced in different countries. The visions and agricultural policies are mainly based on the premise of increasing agricultural productivity to reduce poverty by increasing economic growth. This is being implemented alongside major statements regarding food security. Current existing policies are generally supportive of agricultural practices (example: expansion of cultivated land, increasing farm mechanization, use of organic fertilizer and other inputs) that focus on increasing short term production. But they are less supportive of agricultural practices which improve food production, enhance adaptive capacity and address mitigation (example: restoration of degraded lands, improving soil micro and macro nutrients).

## 6.4 Characteristics of Indian AAS and Exploration of 'Adaptive' Attributes

Over many decades regular use of old ICTs such as television or telephone was rare for large section of the rural population in India. The most remarkable exception was radio, which became widespread quickly due to the availability of battery powered transistorised receivers at cheap price. GKMS have maintained communication with different agricultural departments and produced radio programmes on different agricultural topics on regular basis for broadcasting to rural people through state owned radio stations. Sometimes TV programmes or educational videos were also produced for screening via audio visual vans or on the state run TV stations. Contents were created and controlled by AAS organisations and targeted at the farmer recipient. Since the turn of the millennium, ethereal growth of private mobile phone ownership and use of mobile phone in both rural and urban areas, increasing access to TV, video screening facilities and digital filming apparatus (like mobile phones, cameras) and more recent spread of internet access in cities and even into smaller towns and villages via mobile net services, have offered a new world of opportunity for multi-directional communication system. AAS has accepted new participatory approaches such as farmer field schools and farmer participatory research to mobilise communities and harness complementary contributions from researchers, farmers and AAS staffs for innovation. In general it is seen that AASs are relatively slow to explore new opportunities for a comparable revolution in information sharing, knowledge creation and advocacy activities offered by new and old ICTs. A major challenge for ICTs in AAS and climate change issues is to break the unidirectional communication traditions of the past and develop ICTs as multi way platforms. AAS staffs need information about climate change and its impacts on local environment. Official meteorological stations and researchers are the sources of this information, but both AAS leaders and researchers also need to learn from the experience of farmers, frontline AAS staffs and other sector staffs working at the field level. Adaptive capacity varies widely among individuals and communities due to difference in access to and control of assets and the environments in which people are living. To strengthen adaptive capacity, GKMS needs to recognise these differences and develop strategies to erase them. Self-organisation is a key element of adaptive capacity. Climate change has only emerged as a critical issue and AAS staffs must be received specific training related to climate change in their formal training. This is starting of change, but many AAS staffs have limited their capacity to seek and use new knowledge and information. This is a critical factor that has a major influence on the extent to which AASs access information and networking.

### 6.4.1 AAS advisory methods

In moving towards adaptive AAS, the used advisory methods are critical. To deal with climate change and other uncertainties, AAS methods need to focus on strengthening the capacity of clients rather than delivering messages, enhancing local level innovation, enabling clients to use climate information, strengthening the self-organisation of farmers, improving links between researchers and extension workers, improving the content of advice (eg, innovation and climate resilience).

### 6.4.2 Climate change and AAS initiatives

#### 6.4.2.1 Availability, access and use of climate & weather information

Climate knowledge includes not only climate science, but also local knowledge, interpretations of climate and adaptation practices. Climate knowledge is not uniform and it depends on the clarity of knowledge and the level of vulnerability to the risks involved in climate change trends or other hazards.

- **Low clarity of climate knowledge:** In such conditions emphasis might be given on improving the understanding and increasing the investment in climate modelling or on strengthening the network capacity to access demand to more relevant climate knowledge. Strengthening of adaptive capacity and resilience acts as a buffer to low clarity of climate knowledge.
- **Low vulnerability to a hazard:** This is assessed via a 'starting point' vulnerability analysis. Low vulnerability does not require urgent action but high vulnerability needs urgent action.
- **Higher levels of clarity of climate knowledge, combined with high vulnerability to a particular hazard:** Priority may be given to the implementation of specific adaptation responses (eg, development of drought tolerant crops).

Projections of climate change at country level may vary in quality and coverage. A major reason for this variation is the lack of long term local data sets in detail. Other than climate science, there are some important sources of climate knowledge such as observations by local farmers, indigenous and local adaptive practices used by farmers. Access to this knowledge varies with education, power and influence. Instead of having availability of climate knowledge, there is lack of capacity to use of different types of climate knowledge in adaptation planning. Key questions that AAS can apply to their own organisations in an evaluative sense are the following:

1. What access to climate change science and climate related information do AAS themselves have across the Country?

2. Who are they sharing this information and is it available in a useable form?
3. Are research programmes sharing climate change knowledge and information with the participated farmers? How are they sharing the information and is it being shared in an effective manner?
4. Do AAS have a feedback loop on the impacts of climate change on diverse production systems, local micro climates, markets and livelihoods?
5. How will AAS need to change in relation to climate change information?

Different extension approaches like study circles and farmer field schools bring farmers and rural stakeholders together to discuss about weather, farm operations and livelihoods, but these have to be scaled up and more informed (information on uncertainty and vulnerability) and use of more effective ways of downscaling climate forecast so that forecast information become more useful to specific agro meteorological zones (Christoplos, 2010b). The greatest challenge is supporting smallholders on a sustainable basis in using and adapting to this new type of information on a large scale and in a coordinated fashion. A key factor is the importance of building trust with users through repeat provision of information. Some farmers in India are already collecting and using climate information such as seasonal weather forecast and crop advisories in combination with adaptations to probable market opportunities and risks. However, there are some examples of support for smallholders focused on farm production strategies based on information on seasonal weather forecasts and advice on suitable crops, varieties, farming methods and market probabilities etc., particularly when looking beyond the researcher led pilots and within community based adaptation projects to sustainable provision of services (Christoplos, 2010b).

### 6.4.2.2 AAS and climate change adaptation

There is a wider range of potential adaptation option for making positive changes in existing systems at the farm level.

- Adaptation of conventional, conservation or organic agriculture as farming systems.
- Adaptation of different agricultural practices like soil and water management, crop management, seed management, agro forestry, avoid deforestation and reforestation, pest and disease management.
- Adaptation of livestock and pasture management practices.
- Adaption of farm level climate change mitigation practices.
- On farm and off farm diversification in livelihood practices.
- Diversification of choosing crop species and varieties.

- Formation and evelopment of farmer organisations and social networking.
- Taking up new climate finance, value chain and learning opportunities.

Different experimental evidences show that agricultural or agronomic adaptations have some positive effect in the face of climate change. The diversity in farming environment, the complexity in livelihood strategies of marginal communities (Morton, 2007) and the uncertainties in climate change in combination with other trends, suggest a need of supporting localised innovation to enhance and sustain agricultural performance and resilience. Local innovation processes involve not only the farm level modification and generation of new technologies but also the articulation of demand for testing of the existing technologies that may be appropriate to new conditions. Now it is usually recognised that society has a debt to farmers to pay for those activities (and presumably subsidise). Yet, there is a little consensus about how to undertake such payments on the massive scale, or how to address national and global food security where mitigation measures reduce overall production level. GHG emission has a direct negative impact on agricultural trade.

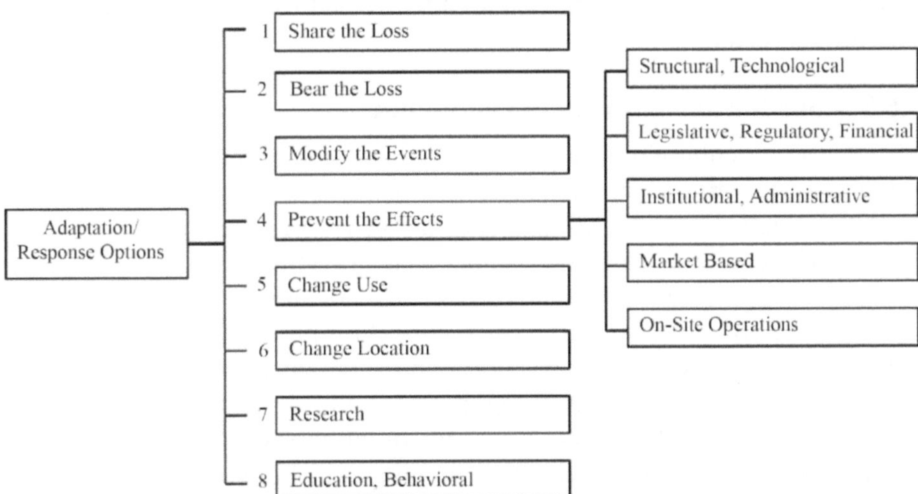


**Fig. 6.4:** Adaptation to climate change and variability. (*Source*: Burton, 1996)

## 6.5 Conclusions and Recommendations

The agricultural context is dynamic, heterogeneous and complex with increasing local, global, social and environmental interactions. In the developing countries agriculture is also a major contributor to the climate change at an increasing fast rate.

To respond to climate change **adaptive AAS** will have to develop in three ways: (i) develop the capacity to manage risk and uncertainty; (ii) recognise and embrace the plurality of AAS; and (iii) response to change and unpredictability. Additionally, the following objectives should also be achieved successfully to support adaptive capacity in rural areas: (i) increasing the availability of key assets for livelihood; (ii) an equitable institutional environment, including attention to gender and social inclusion; (iii) improving the capability to collect, analyse and disseminate the knowledge and information in support of adaptation activities; (iv) achieving an enabling environment fostering innovation, experimentation and niche solutions; and (v) improving the ability to anticipate, incorporate and respond to changes regarding future planning and governance structures.

Though a wide range climate change projections are available, there are some major uncertainties about actual impacts. For AAS to respond to climate change, individuals and organisations need a broad .understanding of climate science, projections and impact models, other forms of climate knowledge, and of vulnerability and adaptive capacity assessments. It is more important that scenario making exercises are undertaken, engaging different sets of stakeholders, to understand the potential future pathways for development of agricultural sectors.

## 6.6 Visions for agricultural development

Visions for agricultural development vary, but there are prevalent notions of agricultural growth and production. Climate resilience and mitigation practices, as well protection services of ecosystem are gaining greater recognition day-by-day. There are several new factors that are influencing investment in agriculture like increasing climate change, concern about food security, rising in food price and volatile energy price, global oil production at its peak and limited availability of natural resources and adaptive ecosystems which can cause serious consequences requiring greater resilience to shocks and stresses. In order to strengthen the adaptive capacity of farmers, GKMS should explore different scenarios with farmers. This requires the ability to (a) identify and analyse the challenges and the opportunities, (b) access information and know how to use them, and (c) put the newly acquired knowledge to use. The ability of AAS individuals and organisations to contribute towards innovation is determined by their internal capacity. In terms of access and use of ICTs in adaptive AAS, the mobile phone revolution is particularly salient for remote and rural communities. There is a significant potential of ICTs to improve access to information related to climate change for interactive communication, networking and shared learning. However, advisory producing methods used by GKMS may need some change for making an effective response to climate change.

## References

Betts R., Gornall J., Wiltshire A. and Climate Change Impacts Team 2009. Impacts of climate change of agriculture. UK Met Office/Hadley Centre.

Boko M., Niang I., Nyong A., Vogel C., Githeko A., Medany M., Osman-Elasha B., Tabo R. and Yanda P. 2007. AfricaIn: Parry ML, Canziani OF, Palutikof JP, van der Linden PJ and Hanson CE eds. Climate change 2007: impacts, adaptation and vulnerability. Contribution of Working Group II to the Fourth Assessment Report of the Intergovernmental Panel on Climate Change. Cambridge University Press, Cambridge UK. p 433–467.

Burton I. 1996. The growth of adaptation capacity: practice and policy. In: Adapting to Climate Change: An International Perspective [Smith, J., N. Bhatti, G. Menzhulin, R. Benioff, M.I. Budyko, M. Campos, B. Jallow, and F. Rijsberman (eds.)]. Springer- Verlag, New York, NY, USA, pp. 55–67.

Christoplos I. 2010a. Climate information and agricultural advisory services: a square peg in Cooke RD. 2010. Investing in agricultural research and agricultural biotechnologies. Paper presented at FAO international technical conference on Agricultural Biotechnologies in Developing Countries (ABDC-10), held in Guadalajara, Mexico, 1–4 March.

Davies M., Oswald K. and Mitchell T. 2009. Climate change adaptation, disaster risk reduction and social protection. In: Climate change adaptation, disaster risk reduction and social protection. OECD. p 201–217.

Davis K.E. 2009. The important role of extension systems. In: Nelson GC ed. Agriculture and climate change: an agenda for negotiation in Copenhagen. IFPRI 2020 Vision Focus 16 Brief 11. International Food Policy Research Institute, Washington, DC, USA. 2 p.

FAO, 2007. Towards an African common market for agricultural products. Report TCP/RAF/3007-II. FAO, Rome, Italy.

Gough C. and Shackley S. 2001. The respectable politics of climate change: the epistemic communities and NGOs. International Affairs. 77(2), 329–346.

Hijioka Y., Lin E., Pereira J.J., Corlett R.T., Cui X., Insarov G.E., Lasco R.D., Lindgren E. and Surjan A. 2014. Asia. In: Climate Change 2014: Impacts, Adaptation, and Vulnerability. Part B: Regional Aspects. Contribution of Working Group II to the Fifth Assessment Report of the Intergovernmental Panel on Climate Change.

Hogan A., Berry H.L., Ng S.P. and Bode A. 2011. Decisions made by farmers that relate to climate change. Rural Industries Research and Development Corporation, Australian Government, Barton, ACT, Australia. 78 p. ISBN 978-1-74254-173-0

IPCC, 2007. Climate change 2007. Synthesis report. Contribution of Working Groups I, II and III to the Fourth Assessment Report of the Intergovernmental Panel on Climate Change. [Core Writing Team, Pachauri RK and Reisinger A eds]. IPCC, Geneva, Switzerland.

Leach M., Scoones, I. and Stirling A. 2010. Dynamic sustainabilities: technology, environment, social justice. Earthscan, London, UK.

Morton J.F. 2007. The impact of climate change on smallholder and subsistence agriculture. PNAS. 104(50), 19,680–19,685.

Nelson V., Lamboll R. and Arendse A. 2008. Climate change adaptation, adaptive capacity and development. Background Paper, DSA-DFID Policy Forum 2008: International Development in the Face of Climate Change: Beyond mainstreaming?

Parry M.L., Canziani O.F., Palutikof J.P. and co-authors 2007. Technical summary. p 23–78. In: Parry ML, Canziani OF, Palutikof JP, van der Linden PJ and Hanson CE eds. Climate change 2007: impacts, adaptation and vulnerability. Contribution of Working Group II to the Fourth Assessment Report of the Intergovernmental Panel on Climate Change. Cambridge University Press, Cambridge, UK.

Smith P., Martino D., Cai Z., Gwary D., Janzen H., Kumar P., McCarl B., Ogle S., O'Mara F., Rice C., Scholes B., Sirotenko O., Howden M., McAllister T., Pan G., Romanenkov V., Schneider U., Towprayoon S., Wattenbach M. and Smith J. 2007. Greenhouse gas mitigation in agriculture. Philosophical Transactions of the Royal Society of London B Biological Sciences. 363(1492), 789–813.

Swanson B.E. 2008. Global review of good agricultural extension and advisory services practices. FAO, Rome, Italy.

Tamiotti L., Olhoff A., Teh R., Sommons B., Kulaoglu V. and Abaza H. 2009. Trade and climate change: a report by the United Nations Environment Programme and the World Trade Organization. WTO, Geneva, Switzerland.

UNEP, 2011. Decoupling natural resource use and environmental impacts from economic growth. A Report of the Working Group on Decoupling to the International Resource Panel. UNFCCC. 2007. Investment and financial flows to address climate change. United Nations Framework Convention on Climate Change, Bonn, Germany. 272 p. ISBN 92-9219-042-3.

# 7

# Contingent Crop Planning in Flood and Drought Conditions of Bihar

## 7.1 Introduction

The occurrence of extreme weather events like drought and flood is increasing in Bihar due to climate change. Bihar receives nearly 80% of annual rainfall during June to September which is quantitatively enough for growing most of the crops. However, aberration in temporal and spatial distribution of rainfall makes the crops vulnerable to drought as well as flood. Rainfall is the most important factor for planning of crop production in rainfed areas. The information on annual, seasonal and weekly rainfall of a region is helpful to design different agricultural operations like field preparation, sowing, irrigation, fertilizer application as well as overall crop planning (Singh *et al.*, 2008). Zone I and zone II of Bihar are receiving too much rainfall which causes flood in this region whereas zone III B is identified as rainfed region which has low, erratic and uncertain rainfall pattern with frequent dry spells during the monsoon season. Hence *kharif* cropping is tricky operation in this region and sudden crop failures is a common phenomenon due to frequent dry spell or early withdrawal of monsoon. In India many scientists studied on rainfall probability pattern (Suchit *et al.*, 2012) and they concluded that rainfall occurrence is certain at greater than or equal to 80 % probability, while 50% probability is the medium limit of certainty and may involve dry spell risk. Based on these climatic and probability factors, the study was carried out for four different locations situated in different agro climatic zones of Bihar for interlinking the rainfall probability pattern with the crop planning pattern in those regions.

Kumar *et al.* (2008) analyzed weekly rainfall data of Bihar for the period of 58 years (1955 to 2012). Data from Pusa, Purnia, Sabour and Patna representing North West alluvial plain (Zone I), North East alluvial plain (Zone II) and South Bihar alluvial plains (Zone III A and III B) of Bihar respectively was used for analysis. Weekly, seasonal and annual rainfall distribution patterns were examined and analyzed. Trends were examined by Mann-Kendall rank statistics as described by Sneyers (1990). Many researchers used this test to detect trends in hydrological time series data (Luo *et al.* 2008). Initial and conditional probability of weekly rainfall at 10, 20 and 30 mm threshold limits were calculated by using first order Markov chain process (Robertson, 1976). Expected amount of rainfall at a given probability level was computed during monsoon season (24-39 Standard Meteorological Week) by using Weibull's distribution (Chow, 1964).

## 7.2. Variability in Annual and Seasonal Rainfall

The average annual rainfall of Pusa (agro climatic zone I) is 1246.9 mm with coefficient of variation (CV) value of 30.8% (Table 7.1) (Kumar *et al.,* 2014). About 83.8% of the annual rainfall is received from south west monsoon. The contribution of pre-monsoon, post-monsoon and winter season rainfall is about 7.9, 5.9 and 2.4% to the annual rainfall respectively. For Purnia (zone II), the average annual rainfall is 1466.7 mm with CV value of 25.8 %. The pre-monsoon, monsoon, post monsoon and winter season rainfall contributes 12, 80.1, 6.5 and 1.4% to the annual rainfall respectively. The average annual rainfall of Sabour (zone III A) is 1231.4 mm with CV value of 24.5 %. The contribution of pre-monsoon, monsoon, post-monsoon and winter season rainfall is 10.4, 78.7, 8.1 and 2.8% respectively. The average annual rainfall of Patna (zone III B) is 1031 mm with CV value of 23.7%. Monsoon season rainfall contributes 85.9% and the pre-monsoon, post-monsoon and winter season rainfall contribution is 4.5, 6.3 and 3.3% respectively.

### 7.2.1 Trend analysis

The long term annual rainfall of Patna (zone III B) shows a significant decreasing trend (Table 7.1) (Kumar *et al.,* 2014). In Pusa (zone I), Purnea (zone II) and Sabour (zone III A), there is statistically not significant increasing trend of long term annual rainfall. There is decreasing trend of winter rainfall in all the four zones, though it is not statistically significant. In Patna (zone III B) there is a significant increasing trend of rainfall in pre monsoon season. Except zone II all the three zones (zone I, IIIA and IIIB) show increasing trend of pre-monsoon rainfall, though it is not statistically significant. But in Pusa (zone I) and Patna (zone III B) there is statistically not significant decreasing trend of monsoon and post-monsoon rainfall.

**Table 7.1:** Variation, percentage contribution (% C) and long term trends of seasonal and annual rainfall in four districts representing different zones of Bihar.

| Season | Pusa (zone I) | | | | Purnia (zone II) | | | |
|---|---|---|---|---|---|---|---|---|
| | Rainfall (mm) | CV (%) | % C | Trend (mm/ year) | Rainfall (mm) | CV (%) | % C | Trend (mm/ year) |
| Annual | 1246.9 | 30.8 | - | -1.57 | 1466.7 | 25.8 | - | 0.81 |
| Winter | 30.3 | 75.8 | 2.4 | -0.48 | 20.6 | 97.5 | 1.4 | -0.10 |
| Pre-monsoon | 98.6 | 59.3 | 7.9 | 1.10 | 175.5 | 54.4 | 12.0 | -0.14 |
| Monsoon | 1044.7 | 34.3 | 83.8 | -1.20 | 1175.3 | 28.3 | 80.1 | 2.26 |
| Post- monsoon | 73.3 | 93.5 | 5.9 | -0.99 | 95.4 | 99.0 | 6.5 | -1.20 |
| | **Sabour (zone III A)** | | | | **Patna (zone III B)** | | | |
| | Rainfall (mm) | CV (%) | % C | Trend (mm/ year) | Rainfall (mm) | CV (%) | % C | Trend (mm/ year) |
| Annual | 1231.4 | 24.5 | - | 2.08 | 1031.0 | 23.7 | - | -7.04* |
| Winter | 34.8 | 89.3 | 2.8 | -0.39 | 34.0 | 79.2 | 3.3 | -0.51 |
| Pre-monsoon | 127.8 | 66.3 | 10.4 | 0.54 | 46.4 | 68.2 | 4.5 | 0.43* |
| Monsoon | 969.7 | 25.0 | 78.7 | 2.39 | 886.1 | 25.1 | 85.9 | -5.74 |
| Post- monsoon | 99.1 | 91.2 | 8.1 | -0.46 | 64.5 | 94.5 | 6.3 | -1.23 |

*Significant at 5% (*Source*: Kumar *et al*., 2014)

## 7.3 Expected Rainfall Amount and Crop Planning

As discussed above, rainfall at 75% and 90% probability is assured rainfall and at 50% probability is the medium limit for taking risk. At Pusa from 25th SMW, there is probability of more than 50% of getting more than 30 mm rainfall, though some risk is there (Kumar *et al.*, 2014). The probability of rainfall occurrence above 15 mm from 27th SMW is more than 75% and farmers can start their field preparation. From 28th SMW expected rainfall becomes above 20 mm, 27th and 28th SMW are ideal time for sowing or transplanting of *kharif* crops and also for fertilizer application based upon the rainfall pattern and intensity. In Purnia from 25th SMW, the probability of getting more than 20 mm rainfall is more than 75 % and the probability of receiving more than 10 mm rainfall is more than 90% (Table 7.2). So, farmers of this region can start sowing of *kharif* crops from 25th SMW. In 28th and 29th SMW probability of getting more than 30 mm rainfall is above 75% which is quantitatively sufficient for transplanting of rice seedlings in this region. In Sabour and Patna region, the probability of receiving more than 10 mm rainfall is above 75% from 25th and 26th SMW respectively (Table 7.3). The probabilities of getting more than 20 mm rainfall in these regions are from 28th and 29th SMW respectively in which transplanting of rice seedlings should be completed.

**Table 7.2:** Expected weekly rainfall (mm) at different probability level (%) in Pusa and Purnia

| SMW | Pusa | | | | Purnia | | | |
|---|---|---|---|---|---|---|---|---|
| | **50 %** | **75 %** | **90%** | **Total** | **50 %** | **75 %** | **90 %** | **Total** |
| 24 | 18.4 | 5.4 | 1.2 | 32.3 | 32.0 | 9.6 | 2.2 | 56.5 |
| 25 | 34.4 | 10.8 | 2.7 | 58.9 | 47.5 | 24.2 | 11.6 | 59.5 |
| 26 | 32.0 | 9.3 | 2.1 | 57.4 | 53.8 | 24.8 | 10.4 | 71.6 |
| 27 | 42.9 | 16.1 | 5.1 | 65.2 | 52.8 | 21.8 | 7.9 | 75.4 |
| 28 | 56.2 | 22.6 | 7.9 | 81.8 | 85.4 | 41.3 | 18.5 | 111.1 |
| 29 | 50.9 | 20.3 | 7.1 | 74.2 | 82.6 | 39.8 | 17.8 | 107.5 |
| 30 | 46.5 | 19.8 | 7.5 | 65.0 | 50.6 | 25.7 | 12.2 | 63.6 |
| 31 | 42.2 | 17.8 | 6.6 | 59.3 | 53.5 | 23 | 8.8 | 74.5 |
| 32 | 49.7 | 21.0 | 7.8 | 70.0 | 37.7 | 17.3 | 7.2 | 50.1 |
| 33 | 47.5 | 19.1 | 6.7 | 69.0 | 42.1 | 14.4 | 4.1 | 67.8 |
| 34 | 47.2 | 19.1 | 6.8 | 68.3 | 45.2 | 20.9 | 8.8 | 60.0 |
| 35 | 33.3 | 12.6 | 4.0 | 50.0 | 34.9 | 13.0 | 4.1 | 53.3 |
| 36 | 38.8 | 16.0 | 5.8 | 55.3 | 48.1 | 24.1 | 11.3 | 60.8 |
| 37 | 39.4 | 15.4 | 5.2 | 58.2 | 51.2 | 22.6 | 9.0 | 69.9 |
| 38 | 23.8 | 8.1 | 2.2 | 38.1 | 29.8 | 11.5 | 3.8 | 43.9 |
| 39 | 26.1 | 6.5 | 1.2 | 51.9 | 46.7 | 17.0 | 5.2 | 72.4 |

(*Source*: Kumar *et al.*, 2014)

**Table 7.3:** Expected weekly rainfall (mm) at different probability level (%) at Sabour and Patna.

| SMW | Sabour | | | | Patna | | | |
|---|---|---|---|---|---|---|---|---|
| | **50 %** | **75 %** | **90 %** | **Total** | **50 %** | **75 %** | **90 %** | **Total** |
| 24 | 22.6 | 7.0 | 1.7 | 38.4 | 14.7 | 3.6 | 0.6 | 29.6 |
| 25 | 31.5 | 12.8 | 4.6 | 45.1 | 24.0 | 8.2 | 2.3 | 38.3 |
| 26 | 39.9 | 16.8 | 6.2 | 56.2 | 31.2 | 10.3 | 2.8 | 51.2 |
| 27 | 34.9 | 12.8 | 4.0 | 53.6 | 50.3 | 20.8 | 7.6 | 71.8 |
| 28 | 54.2 | 22.3 | 8.0 | 77.7 | 47.6 | 17.6 | 5.5 | 73.1 |
| 29 | 45.8 | 19.1 | 7.0 | 64.9 | 72.3 | 33.4 | 14.1 | 96.5 |
| 30 | 33.1 | 12.5 | 4.0 | 49.7 | 49.5 | 18.2 | 5.6 | 76.2 |
| 31 | 40.4 | 16.1 | 5.6 | 58.8 | 47.1 | 22.5 | 9.9 | 61.3 |
| 32 | 35.9 | 13.1 | 4.0 | 55.3 | 49.6 | 22.8 | 9.6 | 66.0 |
| 33 | 42.9 | 15.6 | 4.8 | 66.4 | 40.6 | 15.7 | 5.2 | 60.5 |
| 34 | 26.8 | 8.8 | 2.3 | 44.1 | 55.3 | 26.2 | 11.4 | 72.5 |
| 35 | 23.4 | 7.8 | 2.1 | 38.0 | 33.7 | 13.8 | 5.0 | 48.0 |
| 36 | 25.4 | 8.0 | 2.0 | 43.2 | 36.9 | 15.8 | 6.0 | 51.3 |
| 37 | 23.4 | 8.5 | 2.6 | 36.0 | 44.6 | 17.7 | 6.1 | 65.2 |
| 38 | 20.4 | 5.0 | 0.9 | 41.4 | 16.0 | 5.5 | 1.5 | 25.3 |
| 39 | 24.0 | 5.3 | 0.8 | 52.1 | 25.6 | 7.7 | 1.8 | 44.9 |

(*Source*: Kumar *et al.*, 2014)

## 7.4 Contingent Plan for Flood Affected Area

If due to flood there is total loss of rice seedlings, short duration rice varieties like Prabhat, Rajendra bhagawati, Dhanlaxmi, Turanta, Richhariya, Saket-4 etc. can be sown for new seedling. By dapog method rice seedlings can be raised for transplanting within 10-12 days. Germinated rice seeds can also be sown directly in the field. After flood, 50-60 days old 6-8 rice seedlings per hill can be transplanted. The dose of nitrogenous fertilizer can be increased.

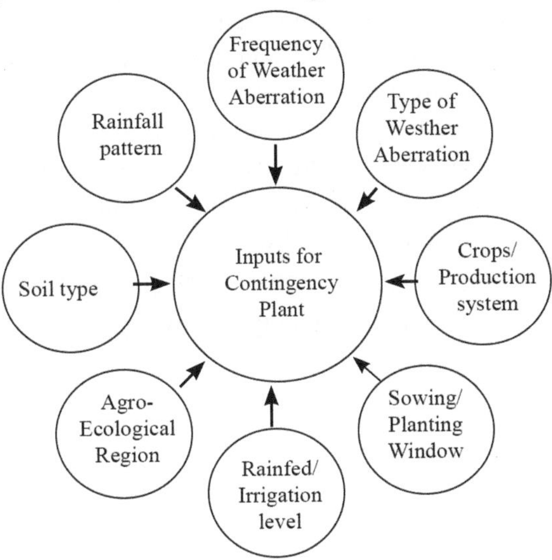


**Fig. 7.1:** Inputs for developing agriculture contingency plan.

## 7.5 Contingent Plan for Drought Affected Area

### 7.5.1 Drought Preventive Measures

Among the different *kharif* crops, upland rice is mostly affected by drought. Therefore, diversified land use with low water requirement non-paddy crops is the best option in drought affected lands. Use of the available technology to mitigate drought requires early planning. The age-old adage "Prevention is better than cure" thus holds good in drought management. Therefore, a long term policy and planning are necessary at the beginning of the crop growing season for judicious use of land and water for getting best results. In rainfed areas for drought mitigation the major emphasis should be given on *in-situ* conservation of rain water, water harvesting through farm reservoirs or capturing runoff water from local catchments or flash flood water to recycle at the time of need. Some important preventive measures that can be adopted early in the crop growing season to mitigate the effect of drought and get sustainable crop production are discussed below.

#### 7.5.1.1 For Upland

Select suitable cropping system and efficient crops matching the length of crop growing season. For rainfed uplands the promising non-rice crops are maize, black gram, red gram, cowpea, sesame, ragi, sweet potato and pumpkin. Select short duration varieties having fast growth rate, deep root system and ability to escape drought condition. Rain water should be stored for using as lifesaving irrigation. On farm water harvesting structures lined with 6:1 soil: cement mortar of 6 cm thickness in 10% land area helps to harvest the rainwater for providing protective irrigation. Off season ploughing will help to conserve soil moisture, reduce weed and pest problem; and also to facilitate early sowing. Follow partial mechanization to ensure timely operations and to utilize land, and other natural resources effectively. Adopt intercropping and mixed cropping system in drought prone areas. Some examples of intercropping and mixed cropping system are mentioned below:

In the areas having rainfall between 625 to 850 mm and medium to deep soil with 200-300 mm water holding capacity, following intercropping system can be suitable.

Maize + Soybean (1:2)

Jowar + Arahar (2:1)

Maize + Groundnut (1:3)

Arahar + Soybean (1:2)

The areas where average annual rainfall is more than 900 mm and there is deep soil with300 mm water holding capacity, following intercropping will be helpful.

Maize + Gram

Maize + Safflower

Soybean + Safflower

Soybean + Gram

Cultivation of cover crops, ridge and furrow method of planting can be done as *in-situ* soil and water conservation method. Use dams, stone structures and nallahs to store rainfall and runoff water. Water harvesting through digging ponds and utilize that harvested water through micro-irrigation methods like drip irrigation and sprinkler irrigation. Mulching will be helpful for moisture conservation. Growing of grasses is also helpful in reducing soil erosion. Construct percolation tanks in light textured soils to recharge the ground water profile and for using as supplementary irrigation. Strengthen village institutions to enable people's participation. Apply FYM at the time of sowing for conservation of soil moisture to prevent seedling mortality due to early drought. Select short duration rice varieties such as Prabhat, Rajendra bhagwati, Dhanlaxmi, Turanta, Richhariya, Saket-4 etc. Sow non-paddy crops like green gram, black gram, black gram, cowpea, maize, sesame, groundnut, castor, ragi in place of rice. Vegetables give good result in drought or low rainfall years. Utilize the reservoirs, ponds and other water bodies for growing chilli, tomato, lady's finger, brinjal, radish, cauliflower, runner bean and cowpea.

### 7.5.1.2 For Medium / low land

Rainwater conservation in medium and low lands is crucial for mitigating drought and improving crop production. A technology has been developed by devoting 10% of the cultivated area for storing excess rain water in medium and low land areas. The objective is to minimize water runoff by increasing infiltration rate (*in-situ* water conservation) and capturing the extra water that cannot infiltrate into the soil. First the field bund is strengthened by raising the height to 45 cm for checking over flow of runoff collecting tank (refuge). The refuge is constructed at the lower reach of the plot with 3.0 m top width, 2.0 m bottom width and 1.8 m depth. The length of the refuge tank must be equal to the width of the plot. This technology is based on the principle that out of total annual rainfall nearly 50% of the rainfall comes from a few intense showers resulting in higher runoff. On the other hand, in certain years during the crop growing period there is a break in rainfall for 10-12 days. Such long dry spell affects the rice crop adversely. In the early season if excess runoff water is collected during intense showers in the refuge, then this collected water can be used as protective irrigation to mitigate the intermittent drought in rice crop.

## 7.5.2 Drought Ameliorative Measures

It is very difficult to define weather scenario during an anticipated drought or dry spell. Hence, it is a difficult task to select contingent measures in advance for the cropping season. Three distinct periods of *kharif* season related to crop growth stages and seven types of scenarios depending on the onset and distribution pattern of rainfall have been projected and required contingent measures have been suggested below.

### 7.5.2.1 Early season drought (June 10 to July 31)

### Scenario 1: Early onset and sudden stoppage of monsoon

Under such situation mortality of sprouts and seedlings is more common problem in sowing.

### 7.5.2.1.1 For Upland

When mortality is more than 50%, re-sowing of the crops should be done up to month of July after receipt of sufficient rain water. It is always wise to raise low water requiring non-paddy crops like ragi, greengram, blackgram, cowpea, sesame and castor. If there is less than 50% mortality, gap filling of the crops will be helpful. Cultivate okra, brinjal, tomato, cowpea, cauliflower or radish wherever possible.

#### 7.5.2.1.2 For Medium and low land

If plant population in rice field is less than 50%, re-sowing the crop should be done. Medium duration varieties are ideal for such situation. Sprouted seeds may be directly broadcasted or fresh seedlings of short duration varieties may be raised for transplanting. The sprouted rice seeds can also be sown in the lines by using seed drill. If the plant population is more than 50%, weeding and adjustment of plant population by removing and redistributing the hills should be done. Raise rice seedling for transplanting at a reliable water source to save time for further delay. In saline soil, use of FYM or green leaf manure, sowing of sprouted seeds, gap filling by clonal propagation give better result.

#### Scenario 2: Late onset of Monsoon

Drought tolerant non-paddy crops like green-gram, black-gram, sesame, cowpea, ragi, guar, and castor can be sown in place of rice. Cowpea may be sown in the first week of August for fodder purpose. Grow sweet potato in the ridges and allow the furrows to conserve rain water. Vegetables like brinjal, chilli. tomato, lady finger cauliflower, radish and cowpea may also be cultivated. Apply full dose of P, K and 30% of N as basal along with well decomposed organic manure. Major emphasis should be given on *in-situ* conservation of rain water and harvesting of excess run-off water for using as life saving irrigation.

### 7.5.2.2 Mid-season drought (August 1 to September 15)

#### Scenario 3: Non-paddy crops affected

Complete hoeing and weeding should be done to provide dust mulch in non-paddy crop fields. To overcome drought situation spray mixture of 2% KCl and 0.1 ppm boron in black-gram. Foliar spray of 2% urea solution at pre-flowering and flowering stage in green-gram and 1% urea solution in brinjal is very effective to control drought. Effective control measures must be taken against mite and mealy bug which are more severe in dry weather. Application of 2% urea solution in late sowing jute gives good result to encourage plant growth. Top dress nitrogen to ginger @ 60 kg/ha and turmeric 30 kg/ha after receipt of rainfall followed by mulching. Use organic mulching to extend the period of moisture availability in the soil. Thin out the crop to the extent of 25% and use those removed plants as fodder or mulch. Close the drainage holes and check the seepage loss in direct sown medium land rice regularly.

### Scenario 4: *Beushaning* of rice delayed

If the seedling is more than 45 days old, do not practice blind cultivation (*beushaning*) in rice. Gap filling using seedlings of same age or clonal tillers should be done to maintain a uniform distribution of plant. Strengthen the field bunds and close the holes to stop seepage loss. Withhold N fertilizer application up to receipt of rainfall.

### Scenario 5: Transplanting of rice delayed

Generally in such cases over aged rice seedlings are used for transplanting. In case of mid-season drought, 45 days old seedlings of medium duration rice varieties and 60-70 days old seedlings of late sown rice varieties can be transplanted without much reduction in yield. Remove the weeds to prevent water loss and follow protection measures to control blast disease in the nursery. Pulverize the main rice field in dry conditions, if it is not ploughed earlier to save time in final pudding. Use tractor or power tiller or tractor mounted rotavator for speedy land preparation and puddling to cover more area with less time. Follow close transplanting by using 5-7 seedlings per hill. Apply 50% of the recommended nitrogen dose at the time of transplanting. Apply lifesaving irrigation to the nursery seedlings to maintain good health.

### Scenario 6: Transplanted rice affected at early vegetative stage

Provide live saving irrigation only at critical stages through harvested rain water. Remove the weeds to prevent water loss, strengthen the field bunds and close the holes to stop seepage loss. Follow control measures against blast disease in the nursery if existing. Withhold N fertilizer application up to receipt of rainfall. Potassic fertilizers can be applied if soil moisture is available, otherwise wait up to receipt of rainfall.

## 7.5.2.3 Late season drought (September 16 to October 31)

### Scenario 7: Medium and low land rice affected at vegetative/ reproductive stage

Late season drought occurs due to early cessation of monsoon rainfall. In this situation protective irrigation should be provide through harvested rain water. Provide lifesaving irrigation only at critical stages (like flowering, grain filling etc.) in alternate furrows in wide spaced crops. Crops like green gram and cowpea may be harvested for green fodder purpose to avoid their failure as grain crops. Under the situation of complete *kharif* crop failure or where land is remaining fallow, sow the pre-*rabi* crops such as blackgram, horse gram, sesame, castor and niger with residual moisture condition in uplands and well drained medium lands.

One of the major crops of this region is rice, which highly depends on monsoon rainfall for preparation of nursery bed, transplanting of seedlings and maintenance of water level in the field. The analysis indicates that there is need for selection of suitable crops according to the probability of getting rainfall with the onset of southwest monsoon in the region. Timely onset of monsoon will favour for selection of long duration and high water demanding crops like rice. In the districts, where moderate rainfall with erratic monsoon behaviour is observed, late onset of monsoon will lead to selection of medium duration and moisture stress/drought tolerant crops like sorghum, finger millet and ragi. Adoption of direct seeded rice can be the best suited example in such situation. Pulses like black gram and green gram, should to be selected as intercropping based on rainfall variability. Analysis reveals that past recorded rainfall data may be useful tool for projection of future rainfall probability.

## 7.6 Implementation Strategies

Extensive planning at the district and state level with overall support and coordination by Government is required for implementation of agriculture contingency plans. Several ministries under Government of India (Fig. 7.2) and also several organizations are involved in implementation of agriculture contingency plans at village level farmers' fields (Fig. 7.3 & 7.4).

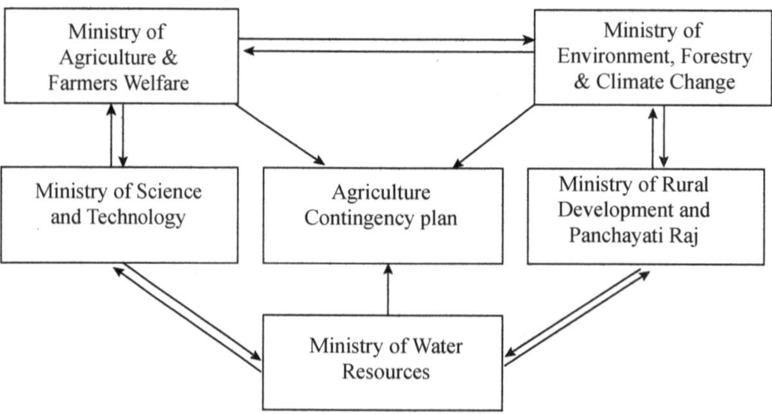


**Fig. 7.2:** Indian Ministries associated with implementation of agriculture contingency plan at village level. (*Source*: Srinivasrao *et al.*, 2020)

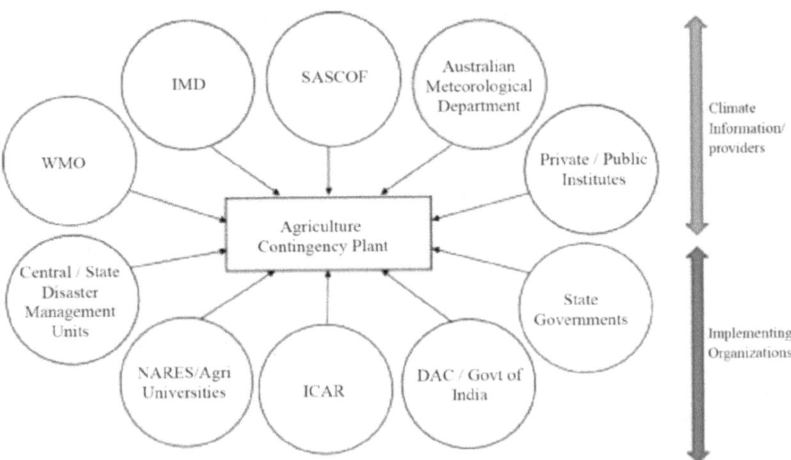


**Fig. 7.3:** Key players for district agriculture contingency plan implementation. (*Source*: Srinivasrao *et al.,* 2020)

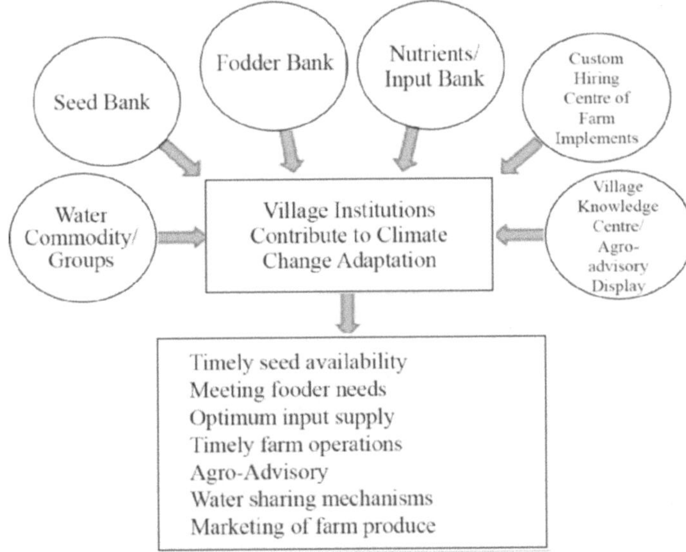


**Fig. 7.4:** Village institutions contribute to climate change adaptation and contingency plan implementation. (*Source*: Srinivasrao *et al.,* 2020)

## References

Ahmed P., Deka R.L., Baruah B.P. and Nath K.K. 2009. Rainfall based crop planning in the Barak valley zone of Assam., J. Agrometeorol. 11(2), 192-195.

Chow V.T. 1964. Statistical and Probability Analysis of Hydrological Data. Hand book of applied Hydrology. V.T. chow editor, McGraw Hill, New York pp 81-89.

Kumar S., Kumar S. and Sharma R.P. 2014. Long-term trends and probability of rainfall for crop planning in Bihar. IndianJ. Ecol. 41(2), 243-246.

Luo Y., Liu S., Fu S., Liu J., Wang G. and Zhou G. 2008. Trends of precipitation in Beijing river basin, Guangdong province, China. Hydrol. Process. 22, 2377-2386.

Robertson, G.W. 1976. Dry and wet spells UNDP/FAO, Ton Rajak Agric. Res. Center, Sungh: Tekam, Malaysia project field report, Agro meteorological, A-6 p.15.

Singh K.A., Sikka A.K. and Rai S. 2008. Rainfall distribution pattern and crop planning at Pusa in Bihar. J.Agrometeorol. 10(2), 198-203.

Sneyers R. 1990. On the statistical Analysis of series of observation.WMO Tech. Note No.143, Geneva.

Srinivasaraoa Ch, Raob K.V., Gopinathb K.A., Prasadc Y.G., Arunachalamd A., Ramanab D.B.V., Ravindra Charyb G., Gangaiahe B., Venkateswarlub B., Mohapatra T. 2020. Agriculture contingency plans for managing weather aberrations and extreme climatic events: Development, implementation and impacts in India. Advances in Agronomy, 159, 36-87.

Suchit K., Behari P., Satyapriya Rai A.K. and Agrawal R.K. 2012. Long term trends in rainfall and its probability for crop planning in two districts of Bundelkhand region. J. Agrometeorol. 14 (1), 74-78.